CMOS SRAM Circuit Design and Parametric Test in Nano-Scaled Technologies

FRONTIERS IN ELECTRONIC TESTING

Consulting Editor
Vishwani D. Agrawal

Books in the series:

Nanometer Technology Designs · High Quality Delay Tests
Tehranipoor, M., Ahmed, N., Vol. 38
ISBN 978-0-387-76486-3
Emerging Nanotechnologies · Test, Defect Tolerance, and Reliability
Tehranipoor, M. (Ed.), Vol. 37
ISBN 978-0-387-74746-0
Oscillation-Based Test in Mixed-Signal Circuits
Huertas Sánchez, G., Vázquez García de la Vega, D. (et al.) , Vol. 36
ISBN: 978-1-4020-5314-6
The Core Test Wrapper Handbook
da Silva, Francisco, McLaurin, Teresa, Waayers, Tom, Vol. 35
ISBN: 0-387-30751-6
Defect-Oriented Testing for Nano-Metric CMOS VLSI Circuits
Sachdev, Manoj, Pineda de Gyvez, José, Vol. 34
ISBN: 978-0-387-46546-3
Digital Timing Measurements – From Scopes and Probes to Timing and Jitter
Maichen, W., Vol. 33
ISBN 0-387-32418-0
Fault-Tolerance Techniques for SRAM-based FPGAs
Kastensmidt, F.L., Carro, L. (et al.), Vol. 32
ISBN 0-387-31068-1
Data Mining and Diagnosing IC Fails
Huisman, L.M., Vol. 31
ISBN 0-387-24993-1
Fault Diagnosis of Analog Integrated Circuits
Kabisatpathy, P., Barua, A. (et al.), Vol. 30
ISBN 0-387-25742-X
Introduction to Advanced System-on-Chip Test Design and Optimi...
Larsson, E., Vol. 29
ISBN: 1-4020-3207-2
Embedded Processor-Based Self-Test
Gizopoulos, D. (et al.), Vol. 28
ISBN: 1-4020-2785-0
Advances in Electronic Testing
Gizopoulos, D. (et al.), Vol. 27
ISBN: 0-387-29408-2
Testing Static Random Access Memories
Hamdioui, S., Vol. 26
ISBN: 1-4020-7752-1
Verification by Error Modeling
Radecka, K. and Zilic, Vol. 25
ISBN: 1-4020-7652-5
Elements of STIL: Principles and Applications of IEEE Std. 1450
Maston, G., Taylor, T. (et al.), Vol. 24
ISBN: 1-4020-7637-1
Fault Injection Techniques and Tools for Embedded Systems Reliability Evaluation
Benso, A., Prinetto, P. (Eds.), Vol. 23
ISBN: 1-4020-7589-8
Power-Constrained Testing of VLSI Circuits
Nicolici, N., Al-Hashimi, B.M., Vol. 22B
ISBN: 1-4020-7235-X
High Performance Memory Testing
Adams, R. Dean, Vol. 22A
ISBN: 1-4020-7255-4

Andrei Pavlov · Manoj Sachdev

CMOS SRAM Circuit Design and Parametric Test in Nano-Scaled Technologies

Process-Aware SRAM Design and Test

Andrei Pavlov
Intel Corporation
2501 NW 229th Street
Hillsboro, OR 97124

Manoj Sachdev
University of Waterloo
Dept. Electrical & Computer Engineering
200 University Ave.
Waterloo ON N2L 3G1

Series Editor

Vishwani Agrawal
Department of Electrical and Computer Engineering
Auburn University
Auburn, AL 36849
USA

ISBN 978-1-4020-8362-4 e-ISBN: 978-1-4020-8363-1

Library of Congress Control Number: 2008924192

springer.com

Pavlov & Sachdev: CMOS SRAM Circuit Design & Parametric Test in Nano-Scaled Technologies

First Indian Reprint 2011

ISBN 978-81-322-0232-5

This edition is published by Springer (India) Private Limited, (a part of Springer Science+Business Media), Registered Office: 3rd Floor, Gandharva Mahavidyalaya, 212, Deen Dayal Upadhyaya Marg, New Delhi – 110 002, India

Printed in India by Rakmo Press (P) Ltd., New Delhi.

To Natasha and Daniel

To Sunanda, Aniruddh and Arushi

Foreword

Embedded SRAMs now dominate CMOS computing chips taking well over half of the total transistor count of high performance ICs. This dominance forces designers to minimize the SRAM layout area imposing a tight transistor density. This transistor circuit density presents two challenges for the test. The first is that virtually all areas of the cells are active and sensitive to particle-related defects. Secondly, parasitic coupling between cells is a major concern. This book addresses both of these problems.

The strongest approach to test method development examines the failure mechanism under test down to the transistor, interconnect and dielectric level. Test detection methods are guided by the electronic properties of the failure mechanism. This analysis style and subsequent development of a test method is called defect-based testing (DBT). This book is a strong example of the DBT thinking. The authors describe an comprehensive SRAM test that is supported by abundant simulation and silicon data. High-density embedded SRAMs require this detailed level of study to understand how to avoid the severe implications of test escapes in this critical majority region of modern ICs.

The authors also supply excellent tutorial descriptions of topics that support SRAMs. This includes design of the memory system and its components, SRAM cell stability, traditional fault models and test practices, I_{DDQ} testing, burn-in challenges and a particularly relevant Chapter 6 on soft error impact on nano-scaled SRAM cells. This book is well written and the reader is the beneficiary of the large amount of work by the authors. The book should be read and on the shelf of engineers who deal with high-performance chips.

Professor Chuck Hawkins
University of New Mexico

Preface

The process technology scaling and push for better performance enabled embedding of millions of Static Random Access Memories (SRAM) cells into contemporary ICs. In several applications, the embedded SRAMs can occupy the majority of the chip area and contain hundreds of millions of transistors. As the process technology continues to scale deeper into the nanometer region, the stability of embedded SRAM cells is a growing concern. As a consequence, large SRAM arrays impact all aspects of chip design and manufacturing because they became the yield-limiters in modern high-performance ICs. However, the robust detection of SRAM cells with marginal stability is a non-trivial task. While the traditional march tests are unable to detect unstable cells, the conventional data retention tests that are intended to expose marginal cells have unacceptable sensitivity and are uneconomical due to the long test time and high-temperature requirements. These factors show the growing disparity between the traditional SRAM test practices and the need for an economical SRAM cell stability tests that can help to achieve lower defect counts in the shipped parts. While various aspects of SRAM design and test have been addressed in special literature, no cohesive text provides a systematic overview of SRAM cell stability and the specialized design and test approaches it requires.

Therefore, the primary objective of this book is to bridge the gap between the challenges that the technology scaling brings on SRAM circuit design and the design and test solutions spanning across the process technology, circuit and system design and the testing. The book gives an overview of SRAM design and the traditional SRAM testing. It builds the reader's understanding on the detailed treatment of the aspects of SRAM cell stability and the state-of-the-art specialized stability testing techniques including the test techniques developed by the authors. In addition, since the unstable SRAM cells are more prone to soft errors, we included an overview of the sources, mechanisms and the mitigation techniques for soft errors in SRAM arrays.

The intended audience of this book is graduate students, engineers and professionals interested in developing understanding and intuition of the challenges faced by modern SRAM design and test engineers.

Authors would like to thank Jose Pineda de Gyvez and Mohamed Azimane (Philips Research Labs) and Patrick van de Steeg (Philips Semiconductors) for several fruitful technical discussions and research cooperation. We appreciate Canadian Microelectronics Corporation (CMC) for providing the chip fabrication services and are grateful to Rutger van Veen and Bram Kruseman (Philips Research Labs) for the help with the creation of the test programs and the test chip measurements. Authors are thankful to Mark de Jongh for facilitating the publishing of this book.

Authors would like to thank their respective families for invaluable support during the preparation of this book. Andrei would like to express special thanks to his wife Natasha for her endless support and encouragement. Manoj would like to express his appreciation to his wife Sunanda, their son Aniruddh and daughter Arushi for their understanding and support.

Andrei Pavlov
Manoj Sachdev

Contents

Acronyms

IC	Integrated Circuit - electronic circuit integrating active and passive components on the same semiconductor substrate
SRAM	Static Random Access Memory - semiconductor memory that is designed to hold the stored data for as long as the power is supplied without the need to refresh it
SNM	Static Noise Margin - a stability metric of an SRAM cell
SoC	System on a Chip - IC that integrates digital, analog, mixed-signal and RF functions on the same chip
ITRS	International Roadmap for Semiconductors - group of experts defining semiconductor technology roadmapping
DRT	Data Retention Test - SRAM test for data retention when an SRAM cell is powered up but not accessed
DRF	Data Retention Fault - failure to retain the data in an SRAM cell when the cell is powered up but not accessed
SF	Stability Fault - failure to retain the data in an SRAM cell in adverse operating/testing conditions
WL	Word Line - access control line of a "word" of SRAM cells
BL,BLB	Bit Line and Bit Line Bar - the true and the complementary input/output lines of a column of SRAM cells
CMOS	Complementary Metal-Oxide-Semiconductor - traditional abbreviation referring to field-effect transistors with metal (or polysilicon) gates and silicon dioxide (or high-k) gate dielectric
DPM	Defect Per Million - measure of fabrication quality of SRAM arrays
IFA	Inductive Fault Analysis - special fault analysis technique that identifies probability of a defect to occur in a given circuit layout
ECC	Error Correction Code - using redundant information to detect and correct data errors
VTC	Voltage Transfer Characteristic - output voltage of a logic gate as a function of input voltage
SER	Soft Error Rate - the rate at which a device or system encounters or is predicted to encounter soft errors. Expressed in FIT or MTBF

FIT	Failure In Time - equivalent to 1 error per billion hours of device operation
MTBF	Mean Time Between Failures - in years of device operation; 1 year MTBF is equal to $\approx$114,155 FIT
PVT	Process, Voltage, Temperature - a set of conditions affecting the circuit
TFT	Thin Film Transistor

Chapter 1
Introduction and Motivation

1.1 Motivation

As the process technology continues to scale, the stability of embedded Static Random Access Memories (SRAMs) is a growing concern in the design and test community [1–3]. Maintaining an acceptable Static Noise Margin (SNM) in embedded SRAMs while scaling the minimum feature sizes and supply voltages of the Systems-on-a-Chip (SoC) becomes increasingly challenging. Modern semiconductor technologies push the physical limits of scaling which results in device patterning challenges and non-uniformity of channel doping. As a result, precise control of the process parameters becomes exceedingly difficult and the increased process variations are translated into a wider distribution of transistor and circuit characteristics.

Large SRAM arrays that are widely used as cache memory in microprocessors and application-specific integrated circuits can occupy a significant portion of the die area. In an attempt to optimize the performance/cost ratio of such chips, designers are faced with a dilemma. Large arrays of fast SRAM help to boost the system performance. However, the area impact of incorporating large SRAM arrays into a chip directly translates into a higher chip cost. Balancing these requirements is driving the effort to minimize the footprint of SRAM cells. As a result, millions of minimum-size SRAM cells are tightly packed making SRAM arrays the densest circuitry on a chip. Such areas on the chip can be especially susceptible and sensitive to manufacturing defects and process variations. International Technology Roadmap for Semiconductors (ITRS) [4, 5] predicted "greater parametric yield loss with respect to noise margins" for high density circuits such as SRAM arrays, which are projected to occupy more than 90% of the SoC area in the next 10 years (Figure 1.1).

As we will discuss in Chapter 3, any asymmetry in the SRAM cell structure, be it due to cell transistor mismatch/offset or due to even subtle process disturbance, will render the affected cells less stable [6]. Under adverse operating conditions such cells may inadvertently flip and corrupt the stored data. Since the bit count of embedded SRAM in microprocessors is constantly growing with every technology generation, the probability of data corruption due to unstable SRAM cells is following the trend.

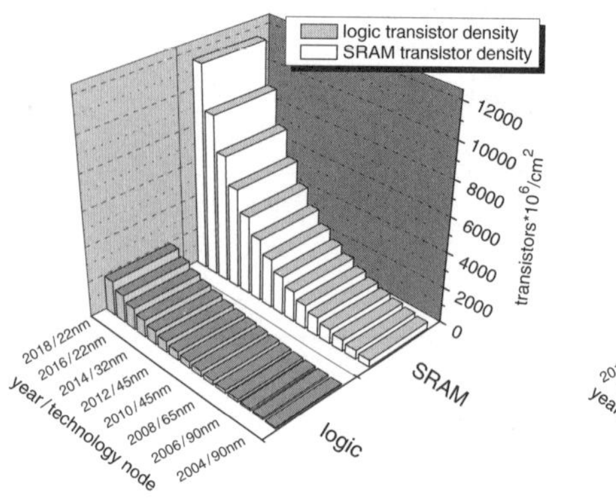

(a) Transistor density trends with scaling: 6T SRAM cell vs. 4T logic gate

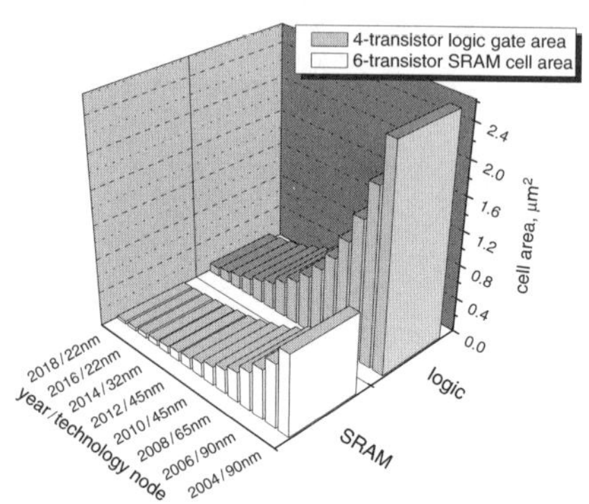

(b) Area trends with scaling: 6T SRAM cell area vs. a 4T logic gate

Fig. 1.1 The predicted high-volume microprocessor technology density and area trends of a six-transistor SRAM cell and a four-transistor logic gate with respect to year/technology node [5]

The SNM can serve as a figure of merit in stability evaluation of SRAM cells. Due to the factors which will be discussed in the following chapters, the SNM of even defect-free cells is declining with the technology scaling. In the presence of even non-catastrophic defects such as poor resistive vias and contacts, and process excursions causing degraded parameters of the PMOS transistors, SRAM cell SNM is degraded even further. SRAM cells with compromised stability can limit the reliability of on-chip data storage making it more sensitive to transistor parameter shift with aging, voltage fluctuations and ionizing radiation. Detection and correction/repair of such cells in modern scaled-down SRAMs becomes a necessity. However, the traditional SRAM tests, such as march test, are incapable to reliably detect such unstable cells.

In this book, we attempted to systematically address the root causes of SRAM cell instability in scaled-down technologies as well as the design and test solutions addressing this problem. We studied the sensitivity of SRAM static noise margin to multiple process and environmental parameters and developed an SRAM cell stability fault model and the stability testing concept. The results of this study presented in Chapter 3 are intended to clarify the principles behind the state-of-the-art SRAM stability test techniques.

Detection of such Data Retention Faults (DRF) caused by unstable SRAM cells has been a time consuming and expensive effort. Traditional Data Retention (sometimes also called Delay or Pause) Test (DRT) has a limited fault coverage with respect to the weak cell detection and is time consuming and may require elevated temperatures. For most of the DRFs and Stability Faults (SFs), the DRT can be ineffective. If left undetected, the stability faults in SRAM cells may lead to circuit/system failures in the field. Since the stability faults are mostly caused by manufacturing defects, they can also indicate potential long-term reliability issues in the product. Since the traditional testing may not detect all of cells with potential stability problems, researchers proposed a number of specialized stability testing techniques that help to ensure lower defect count and the improved quality of the shipped products. These techniques are designed to expose the unstable SRAM cells that would otherwise be undetected. Many of these techniques are discussed in detail in Chapter 5 that gives an overview of the state of the art in SRAM cell stability testing. The results of SRAM stability testing could be used to influence the decisions regarding the repair strategy and the allocation of redundant elements in SRAM arrays as well as performance binning of the chips.

1.2 SRAM in the Computer Memory Hierarchy

Memory has been the driving force behind the rapid development of CMOS technology we have been witnessing in the past few decades. Starting from the first 1 Kb DRAM chip developed by Intel in the 1970s, nowadays DRAM capacities have reached beyond 1 Gb.

The advent of the virtual memory in personal computers contributed to the hierarchical structure of various kinds of memory ranging from the small capacity, fast but more costly cache memories to large capacity, slower but more affordable magnetic and optical storage. The pyramid-like hierarchy of memory types in a Personal Computer (PC), shown in Figure 1.2, reflects the growing speed and cost/bit as we move from the bottom Level 5 (L5) remote secondary storage to the topmost register storage Level (L0) in the PC memory hierarchy. The introduction of memory hierarchy is a fundamental consequence of maintaining the random access memory abstraction and practical limits on the cost and the power consumption.

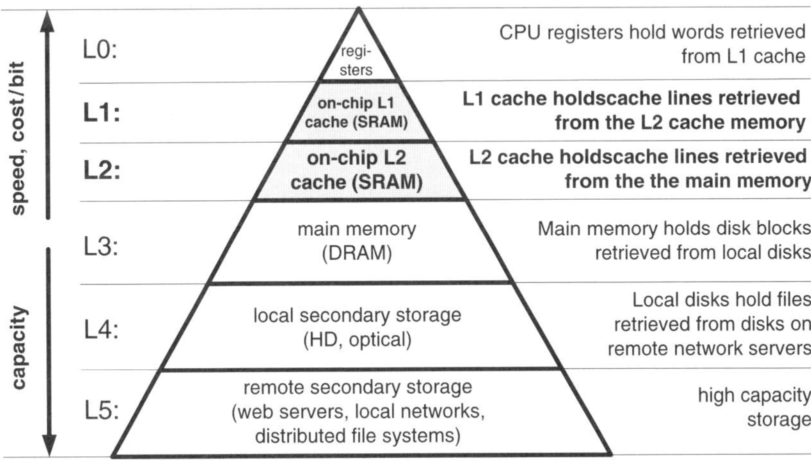

Fig. 1.2 Memory hierarchy of a personal computer

The growing gap between the Micro Processor Unit (MPU) cycle time and DRAM access time necessitated the introduction of several levels of caching in modern data processors. In personal computer MPUs such levels are often represented by L1 and L2 on-chip embedded SRAM cache memories. As the speed gap between MPU, memory and mass storage continues to widen, deeper memory hierarchies have been introduced in high-end server microprocessors [7].

ITRS distinguishes to types of MPU depending on the amount of L2 cache. The first type is the Cost-Performance MPU, which is optimized for maximum performance, and the lowest cost by limiting the amount of on-chip SRAM Level-2 (L2) cache. The second type is the high-performance MPU optimized for maximum system performance by combining a single or multiple CPU cores with a large L2 and recently, L3 on-chip SRAM cache [7]. Logic functionality and L2 cache capacity typically doubles every technology generation by doubling the number of on-chip CPU cores and associated memory.

One way to increase the on-chip cache sizes uses the high-density dynamic RAM. An SoC with embedded DRAMs implemented in the standard logic process can benefit from fast low-V_{TH} transistors. However, the inherently high subthreshold leakage current complicates implementation of a 1T DRAM cell. Replacing 1T DRAM cells with alternative DRAM cell designs having a larger number of transistors results in an area penalty and undermines the cell area advantage that embedded DRAMs normally have over embedded SRAMs. If the DRAM process is used to fabricate embedded DRAM, a 1T DRAM cell can achieve high packing density. However, the high-V_{TH} low-leakage DRAM process limits the performance of such an SOC [8, 9]. Therefore, usage of embedded DRAMs may be justified in specialized SoC, requiring large embedded memory size and operating at relatively low to medium speeds.

On the other hand, embedded SRAMs have successfully accelerated the performance of high-end microprocessors, network routers and switches. They use the regular fast logic process and do not require additional mask steps.

1.3 Technology Scaling and SRAM Design and Test

Ever since the early days of semiconductor electronics, there has been a desire to miniaturize the components, improve their reliability and reduce the weight of the system. All of these goals can be achieved by integrating more components on the same die to include increasingly complex electronic functions on a limited area with minimum weight. Another important factor of successful proliferation of integration is the reduced system cost and improved performance. While the initial drive for miniaturization and integration came from the needs of the military, it quickly proved vital in consumer products as well.

1.3.1 Moore's Law

In 1965 Gordon Moore, the future co-founder of Intel Corporation, showed that one of the major benefits of integration and technology scaling is the reduction of cost per function of the system. He demonstrated that for simple circuits, the cost per component is nearly inversely proportional to the number of components. But as the number of integrated components is growing, the yield starts to deteriorate, tending in turn to raise the cost per component. These trends lead to a minimum cost at any given time in the evolution of the technology. Figure 1.3(a) illustrates these trends for some early semiconductor processes. The relative manufacturing cost per component is thus decreasing with each technology generation, while the number of successfully integrated components is steadily increasing.

Moore made an important observation. Looking at the complexity for minimum component cost in Figure 1.3(a), he noted that the number of the components "increased at a rate of roughly a factor of two per year" [10]. He speculated that this growth would continue over the longer term, although he acknowledged that the "rate of increase is a bit more uncertain." Still, he asserted that "there is no reason to believe it will not remain nearly constant for at least 10 years." After a decade had passed, Moore revisited his law and adjusted his prediction based on the data shown in Figure 1.3(b). He concluded that integrated circuit complexity was still increasing exponentially but at a slightly slower pace. In this revision, he predicted a change of the slope of the curve in Figure 1.3(b) and the slowing down of the growth of the components per chip "to doubling every two years, rather than every year" [11] by the end of the 1970s. The current pace of "cramming more components onto integrated circuits" sets the increase of the number of elements to double every 18 months.

Thc industry has bccn ablc to achicvc thc targcts sct by Moorc's law only thanks to careful and dedicated planning headed by Sematech. In 1991, a consortium of

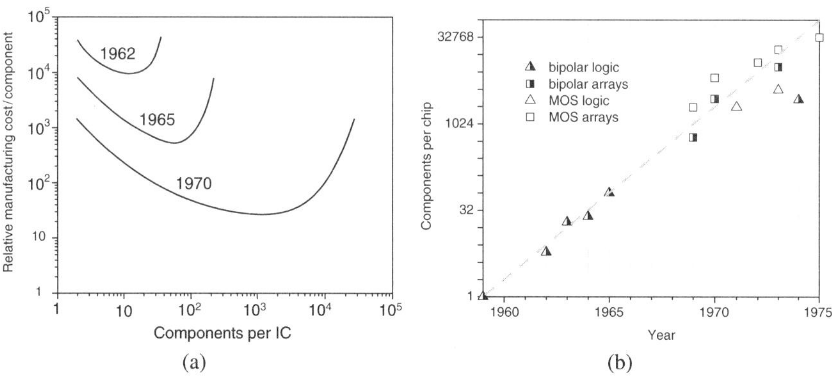

Fig. 1.3 The trend of annual doubling of the number of integrated components at the optimal cost/component that led to the discovery of Moore's law (a) [10]. The prediction of the number of components per chip was corrected 10 years later to doubling every 2 years (b) [11]

global industry manufacturers, suppliers, government organizations, consortia, and universities and the National Advisory Committee on Semiconductors named Sematech (later, International Sematech) started the technology roadmapping process called The International Technology Roadmap for Semiconductors (ITRS). ITRS is an assessment of the semiconductor industry's technology requirements. It suggests time tables to keep the industry on its historic productivity curve [12]. By enshrining it in a major planning document, the Roadmap turned Moores law from a prediction for the future into a commitment for the semiconductor industry. The objective of the ITRS is to ensure advancements in the performance of integrated circuits and remove roadblocks to the continuation of Moore's Law. Continuing Moore's law, however, is far from automatic. "The semiconductor industry's future success," argued the 2001 Roadmap, "continues to depend upon new ideas," ideas to solve problems in lithography, design, testing, manufacturing, structure, assembly, and packages [12].

To double on-chip functionality every two years according to Moore's Law, technology-node scaling of 0.7 in linear size and 0.5 in area has to be carried out every 3 years; as well as an additional device/process design improvement of 0.8/(2 years) must be achieved. The advancement rate through the technology nodes is determined by the ability to economically manufacture a chip based upon the best available leading-edge design and manufacturing process. The typical high-performance ASIC design is assumed to have the same average transistor density as high-performance MPUs, which mostly consist of SRAM transistors [5].

The gap in transistor density between the regular logic and embedded SRAMs is predicted to grow from approximately 5X in 2005 and to around 6X by 2018 for every technology node, as shown in Figure 1.1(a). At the same time, the area of a six-transistor SRAM cell now constitutes only $\approx$36% of the typical four-transistor logic gate area and is projected to reduce further to become $\approx$35% in year 2018 (Figure 1.1(b)). As a rule, SRAM cell size continues to scale $\approx$0.5X/generation driven by the need for high-performance processors with larger caches.

Scaling of a memory cell leads to the increase in the critical area where a spot defect of a given size may cause a bridge or a break in the topology and damage the cell. Combined with the exponentially growing bit count of embedded SRAMs, the probability of a defect or a large parameter variation in an SRAM cell is constantly growing. Therefore, design and testing of embedded SRAMs are becoming more challenging with each technology generation. Special test methods are required to provide efficient and cost-effective testing for reliable screening and/or repairing of the defective and unstable SRAM cells.

1.3.2 Obstacles in SRAM Scaling

There are, however, obstacles on the way of continuous scaling of SRAM. One of them is that SRAM cell power delay-area product is not scaling as efficiently as that of logic circuits. This phenomenon is known as the non-scaling problem of

SRAM [13]. Multiple effects of device dimensions scaling over technology generations yielded faster transistors. However, the saturation current of the scaled transistors often remained on the previous level. The read current of an SRAM cell I_{read} depends on the saturation current of the access transistor and the speed often depends on the small-signal slew rate as:

$$t_{\Delta V_{BL}} = \frac{C_{BL}\Delta V_{BL}}{I_{read}}. \tag{1.1}$$

Equation 1.1 shows that if I_{read} remains at the previous level, then SRAM speed can be improved by reducing the bit line differential developed during a read operation ΔV_{BL} and/or the capacitance of the bit lines C_{BL}. The bit line capacitance C_{BL}, which consists of the diffusion and wire capacitance components, only partially benefits from scaling of the transistor dimensions. The wire capacitance component that is responsible for at least a half of the total bit line capacitance C_{BL} is not scaling down at the same rate as the speed of the logic circuits. The voltage differential required on the bit lines for reliable sensing ΔV_{BL} is not scaling at the same rate as the logic speed. Moreover, the ΔV_{BL} may not scale down and can even increase with the process technology scaling to allow for larger tolerances and margins due to the increasing process variability and difficulty of parameter matching in nano-scaled technologies, which also tend not to scale down.

While the non-scaling problem can be addressed by the modifications to the SRAM architecture, such modifications often involve certain tradeoffs. The extra logic and bypass paths that support these architectural workarounds come at a cost of extra chip area and power dissipation. Another approach reduces the bit line capacitance C_{BL} seen by any given cell by reducing the number of cells per a bit line or introducing a hierarchical bit line structure with global and local bit lines. However, the area and power penalty paid for applying this strategy has reached its practical limits. Other techniques addressing the non-scaling problem that are also reaching their practical limits are multi-stage or complex sense amplifiers, local amplification of the bit line differential ΔV_{BL} as well as aggressive clocking and pipelining methods.

Thus, SRAM non-scaling problem presents one of the most difficult tasks in designing of nano-scaled SRAMs. The possible solutions helping to mitigate the SRAM non-scaling problem are driven by the target application of SRAM arrays with the high-performance applications on one end of the spectrum and the low-power mobile applications on the other.

1.4 SRAM Test Economics

While the manufacturing costs can be decreased by integrating more devices on the same die area, the cost of testing does not scale. Scaling of transistor dimensions, higher packing density and growing defect density contribute to the growing testing complexity with every new technology generation.

Memory testing has become more difficult over the past decades, while parametric faults are becoming more common leading to yield loss and reliability problems. While the number of I/O pins on a chip has increased by one order of magnitude, the number of transistors has increased by up to four orders of magnitude [14]. The increasing transistor/pin ratio which is projected to exceed 2.3 million/pin by 2016 [4] is limiting the controllability from the primary inputs and the observability of the faulty behavior at the primary outputs in embedded memories [15]. Striving to keep up with the increasing clock speeds of SoCs increases the cost of Automated Test Equipment (ATE) so that the at-speed test of high-performance chips becomes problematic. The fastest available ATE is always slower than the chips it tests. In other words, we are testing the tomorrow's technology chips with the today's technology testers. As multi-million-dollar ATEs become commonplace, the cost of the tester time that is spent on the testing of every chip has a direct impact on the total cost of the chip.

A customer regards a product to be of high quality if the product is meeting their requirements at the lowest possible cost. Memory tests check conformance to the requirements, while the cost is reduced by improving the process yield. Quality can be expressed as the number of customer returns per million or Parts Per Million (PPM):

$$Defect\ level = \frac{test\ escapes}{total\ number\ of\ shipped\ chips}\ (PPM) \tag{1.2}$$

It is easy to see that a test with a 99% fault coverage will result in the return level of 10,000 PPM, which is a long way from the industry's aim of sub-100 PPM [16].

Exhaustive functional memory testing has long become economically unfeasible. For instance, an exhaustive full functional test of a 1 Kb SRAM array will take 2^{1024} combinations to complete a full functional test. Such a test of an SRAM with the access time of 10 ns will conclude in more than 10^{290} years! Attention was given to the fault modeling, structural testing and DFT techniques to ensure and maintain test cost effectiveness and low defect levels.

The test cost per chip, which can run up to a half of the product cost and is directly related to the test time, cannot be allowed to increase significantly. However, the number of bits per chip is exponentially growing and the fault sensitivity is increasing. Maintaining an acceptable defect level and low PPM in the upcoming scaled down generations will likely require more complicated, lengthy and hence expensive tests.

1.5 SRAM Design and Test Tradeoffs

Many factors that affect the product quality and cost can impose contradictory requirements on SRAM design and test. Lower product cost calls for packing the maximum number of minimum-size transistors on the minimum chip area with subsequent inexpensive testing. However, certain performance and quality metrics are impossible to meet without using larger transistors and additional circuitry that may not be directly related to the chip functionality. For instance, such features as the BIST blocks for at-speed testing and accessability, error correction circuitry for fault tolerance, redundancy for defect repair and design-for-test circuitry for special testing help to improve the product quality. At the same time, these same features in certain applications can be considered as cost-ineffective or even cost-prohibitive additions that add excessive area and testing time and complexity overhead. Design and test engineers are faced with complex tradeoff decisions that have to be reevaluated for every new product. A careful analysis of all aspects of the product life from design to the application is necessary to strike a right balance between the quality and the cost.

1.5.1 Area and Stability

The area of an SRAM array is largely defined by the area of the cells and the efficiency of the array set as a portion of the area occupied by the SRAM array in the total area of an SRAM block. Both the array area and the block area depend on multiple factors. As we will demonstrate in this book, smaller SRAM cells, while more area efficient, inherently are less stable and less reliable. Large SRAM arrays built on smaller cells may require additional means such as error correction circuitry and redundant elements to achieve higher quality levels. Therefore, the area gain from using smaller SRAM cells for the array can be somewhat offset by the area impact of the additional circuitry on the total area of an SRAM block. Larger SRAM cells that are properly designed tend to provide a more robust data storage elements that are more tolerant to subtle defects and unfavorable environmental conditions such as reduced supply voltage and high temperature.

1.5.2 Quality and Yield

As the technology scales down, the reliable patterning and fabrication of the shrinking device features are becoming more and more difficult. Scrupulous Optical Proximity Correction (OPC) of the drawn layout to improve its repeatable manufacturability and precise control of the fabrication process are essential for limiting the defect density. The defect density is typically not scaling down. This can lead to a larger number of marginally formed or even defective device and interconnect locations on the die. Moreover, as the number of SRAM bits grows, the probability of defects crawling into the larger SRAM array also grows. These factors can adversely impact the yield of SoCs that contain large number of SRAM cells.

Tests have limited defect coverage and differ in terms of the efficiency with respect to various kinds of defects. Achieving high quality that would guarantee a low DPM in the shipped SoCs while maintaining cost efficiency of the test requires tests capable of detecting as many defects as possible in the shortest possible time. Obviously, these two requirements contradict each other. Quality and the test cost are often traded off for each other depending on the marketing requirements of the target application. Moreover, overly stringent testing can by itself introduce the yield loss. This is especially true for parametric and performance testing. For instance, an SRAM array can be fully functional at the full power supply voltage but become unstable and cause a corruption of several data bits at a reduced supply voltage. The decision whether or not such a chip is to be deemed defective and discarded or merely be marketed for a different application has to be made for every type of a product. For instance, a toy or a cell phone controlled by a microprocessor that contains SRAM would be having relaxed DPM requirements. However, mission critical applications such as military, avionics, bank servers, medical, would demand much higher quality and reliability and thus more scrupulous and expensive testing. Apparently, the yield of such mission-critical chips will be traded off for the higher quality.

1.5.3 Test Coverage and Test Time

Table 1.1 presents the test time required to test a memory using march algorithms of various complexity, where N is the number of the tested memory locations. The access cycle time is assumed to be 10 ns.

As apparent from Table 1.1, early memory test methods with test times proportional to $t_{cycle} * Nlog_2N$ or even $t_{cycle} * N^2$ are now prohibitively expensive [14]. The $6N$ and $11N$ tests are considered a better tradeoff between the test time and the test coverage. However, a single march test such as $11N$ cannot alone guarantee a low DPM level in the tested arrays. A suite of several tests that are designed to target the most probable defects in the array are often required to improve the test coverage at the expense of the test time. The increase in the test time is directly translated to a reduced tester throughput and drives up the testing-related portion of the product cost.

An ideal test algorithm(s) should have maximum fault coverage with minimum complexity, which is proportional to the test time and test cost. However, real test algorithms have limited fault coverage. To improve the fault coverage, several tests

Table 1.1 Test time as a function of the memory size (calculated for $t_{cycle} = 10$ ns). Test algorithms of more than linear complexity are not economical for large memories

Size	N (sec)	$6N$ (sec)	$11N$ (sec)	$Nlog_2N$ (sec)	$N^{1.5}$	N^2
1 Mb	0.010	0.063	0.115	0.210	10.7 sec	1.3 days
4 Mb	0.042	0.252	0.461	0.923	85.9 sec	20.4 days
16 Mb	0.168	1.007	1.8	4.0	19 min	325.8 days
64 Mb	0.671	4.0	7.4	17.5	1.53 h	14.3 years
256 Mb	2.7	16.1	29.5	75.2	12.2 h	228.5 years

that will cover the target faults may have to be employed. Therefore, when considering the appropriate testing for each type of product, a test engineer is faced with a difficult choice between balancing the testing cost and the quality level this testing should provide.

1.6 Redundancy

By embedding SRAMs within a SoC manufacturers are able to meet the bandwidth, performance, area and performance/watt ratio more effectively than using stand-alone memory. Embedded SRAM footprint is continuing to grow and often dominates the chip area. However, the shrinking device dimensions and increasing density affect the yield of large SRAMs. Combined with the significant chip area occupied by the embedded SRAMs, the chip yield is often limited by that of the embedded SRAMs. The limited yield, in turn, drives up the fabrication cost of the chip. The obvious solution is to find a way to improve the yield of the embedded SRAMs. One of the effective techniques serving this purpose is enabling the redundancy with the test and repair capability.

Redundancy can be used to repair a certain number of failing cells by replacing the faulty rows and/or columns or single faulty bits with the spare ones. Applying row or column redundancy inevitably introduces a performance penalty for the affected row or column. The bit redundancy when only one faulty bit is replaced by a spare bit incurs a speed penalty only for that one bit. The repair capability can improve the overall manufacturing yield for large SRAM arrays. The exact redundancy yield benefit is determined by a complex relationship between the chip size, SRAM array real estate share and the maturity of the technology.

One of the main objectives of any chip design is to minimize the cost. Two major factors of the chip cost are the number of good chips per wafer (GCPW) and the test cost per chip (TCPC).

Decision on whether to use redundancy is affected by several considerations and represents a trade-off between the cost and the benefits of using the redundancy [17]:

- Smaller chip size can increase the number of chips per wafer (CPW).
- Large SRAM arrays typically have lower manufacturing yield especially for the new, not matured technologies, decreasing GCPW.
- Redundancy allows some of the failing chips to be repaired, increasing the manufacturing yield and GCPW, especially for large SRAM arrays.
- Redundancy requires extra area (redundant rows/columns, fuses and larger BIST controllers), decreasing the CPW.
- Redundancy requires extra test time for registering the failing cell addresses during test, address rerouting by burning the fuses, which increases the TCPC.

There are practical limitations with respect to the amount of non-redundant memory that can be used on a chip. Any SRAM array of a size larger than 256 Kb is suggested to have a repair capability by using redundancy [18]. To maintain an acceptable yield for a larger number of bits in an SRAM array, redundancy becomes indispensable. Since a chip can be rendered faulty by any single failing memory bit, the yield is determined by the total amount of embedded memory, not only by the largest SRAM instance.

During the initial manufacturing test, a map of faulty locations is stored using fuses. Memory arrays with redundancy are connected to a fuse decompression circuitry that reads the fuse values and decompresses them to the fuse shift-register chain under the control of the BIST circuit. During a power-on-reset sequence, the fuse information is decompressed and shifted through the fuse shift-register chain to the corresponding row and column decoders. Based on the fuse information, the decoders reroute the defective address values to the spare rows and columns. Once all the memories in SoC received the redundancy information, a "ready" flag is issued and the SoC can begin to operate. A separate scan chain restores the fuse information on power-up from the sleep mode, which adds to design complexity and requires co-design of SRAM and the BIST for each instance.

Before redundancy can be applied, all the faulty memory locations must be reliably identified. The repeatable or "hard" faults in SRAMs are easily detectable with the traditional march tests. However, detecting retention and/or stability problems in SRAM cells is much more challenging. The DRT that has traditionally been used for the detection of cell with poor data retention will typically add at least 200 ms for each SRAM array (Figure 5.4). Furthermore, the DRT often requires higher temperatures to improve the defect coverage. Often, the resulting impact of using the DRT on the test is cost-prohibitive. Moreover, the DRT is shown inadequate for detection of more subtle defects in SRAM cells.

Many of the DFT techniques presented in this book in Chapter 5 can successfully replace the DRT. These DFT techniques exceed the DRT in test time and defect coverage. Moreover, the flexibility of the pass/fail threshold provided by the programmability of the applied test stress allows control over the balance between the quality and the yield.

Chapter 2
SRAM Circuit Design and Operation

2.1 Introduction

A significantly large segment of modern SoCs is occupied by SRAMs. For instance, SRAM-based caches occupy more than 90% of 1.72 billion transistors in the Montecito processor [19]. Similarly, SRAM content in ASIC domain is also increasing. Therefore, understanding SRAM design and operation is crucial for enhancing various aspects of chip design and manufacturing.

In subsequent sections we will discuss the salient design and operational issues of SRAMs in general and the SRAM cell in particular. SRAM cell design considerations are important for a number of reasons. Firstly, the design of an SRAM cell is key to ensure stable and robust SRAM operation. Secondly, owing to continuous drive to enhance the on-chip storage capacity, the SRAM designers are motivated to increase the packing density. Therefore, an SRAM cell must be as small as possible while meeting the stability, speed, power and yield constraints. Near minimum-size cell transistors exhibit higher susceptibility with respect to process variations. Thirdly, the cell layout largely determines the SRAM critical area, which is the chip yield limiter. Meeting the design constraints requires deeper understanding of the involved trade-offs. In scaled technologies the cell stability is of paramount significance. Static Noise Margin (SNM) of a cell is a measure of its stability. Sections 3.2, 3.3 and 3.4 present an in-depth discussion on SNM and analytical approaches for its computation.

2.2 SRAM Block Structure

Figure 2.1 shows an example of the basic SRAM block structure. A row decoder gated by appropriate timing block signal decodes X row address bits and selects one of the word lines WL_0–WL_N-1. The SRAM core consists of a number of arrays of NxM, where N is the number of rows and M is the number of bits. If an SRAM

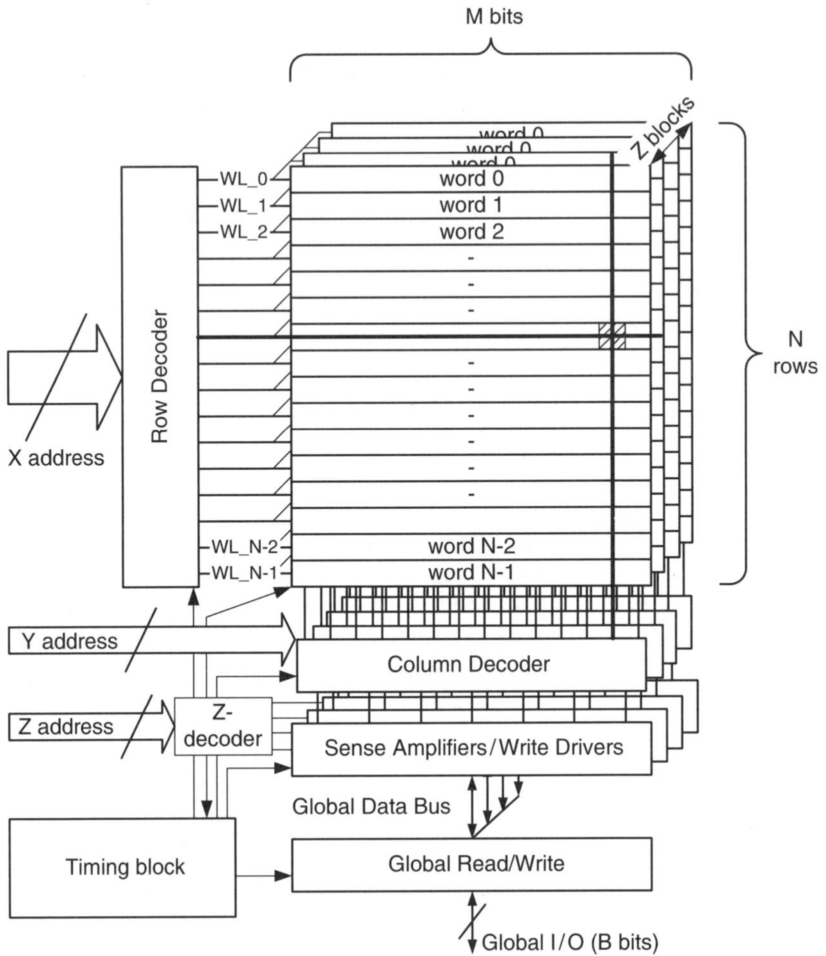

Fig. 2.1 SRAM block diagram

core is organized as a number of arrays in a page manner, an additional Z-decoder is needed to select the accessed page. Figure 2.1 shows an example of an SRAM with four pages of NxM arrays with the corresponding I/O blocks.

SRAMs can be organized as bit-oriented or word-oriented. In a bit-oriented SRAM, each address accesses a single bit, whereas in a word-oriented memory, each address addresses a word of n bits (where the popular values of n include 8, 16, 32 or 64). Column decoders or column MUXs (YMUXs) addressed by Y address bits allow sharing of a single sense amplifier among 2, 4 or more columns. The majority of modern SRAMs are self-timed, i.e. all the internal timing is generated by the timing block within an SRAM instance. An additional Chip Select (CS) signal, introducing an extra decoding hierarchy level, is often provided in multi-SRAM chip architectures.

The main SRAM building blocks will be described in the following sections.

2.3 SRAM Cell Design

An SRAM cell is the key SRAM component storing binary information. A typical SRAM cell uses two cross-coupled inverters forming a latch and access transistors. Access transistors enable access to the cell during read and write operations and provide cell isolation during the not-accessed state. An SRAM cell is designed to provide non-destructive read access, write capability and data storage (or data retention) for as long as cell is powered.

We will discuss design and analysis aspects of three different SRAM cells: a resistive load four-transistor (4T) SRAM cell, a six-transistor (6T) CMOS SRAM cell and a loadless 4T SRAM cell. We will describe their advantages and disadvantages.

In general, the cell design must strike a balance between cell area, robustness, speed, leakage and yield. Cell size minimization is one of the most important design objectives. A smaller cell allows the number of bits per unit area to be increased and thus, decreases cost per bit. Reduced cell area can indirectly improve the speed and power consumption due to the reduction of the associated cell capacitances. Smaller cells result in a smaller array area and hence smaller bit line and word line capacitances, which in turn helps to improve the access speed performance. Reducing the transistor dimensions is the most effective means to achieve a smaller cell area. However, the transistor dimensions cannot be reduced indefinitely without compromising the other parameters. For instance, smaller transistors can compromise the cell stability. Often, performance and stability objectives restrict arbitrary reduction in cell transistor sizes. Similarly, cell area can be traded off for special features such as an improved radiation hardening or multi-port cell access.

2.3.1 *Four-Transistor (4T) SRAM Cell with Polysilicon Resistor Load*

Historically, 4T polysilicon resistor load cells are the remnants of the pre-CMOS technologies. Ratioed inverters in the cell have lower gain in the transition region and produce inherently less steep Voltage Transfer Characteristics (VTCs), which reflects on the SNM values and the time necessary to recover from the metastable state [20].

The main advantage of static 4T cells with polysilicon resistor load (PRL) (Figure 2.2) is the approximately 30% smaller area as compared to 6T CMOS SRAM cells. Due to the higher electron mobility ($\mu_n/\mu_p = 1.5-3$), all transistors in a PRL cell are normally NMOS. The load resistors serve to compensate for the off-state leakage of the pull-down devices. On one hand, the values of R_L must be as high as possible to retain a reasonable noise margin NM_L, i.e., to limit the "0" level rise and reduce the static power consumption. On the other hand, a high R_L severely increases the low-to-high propagation delay for $V_{DD}/2$ precharge and it also increases the cell size. $V_{DD}/2$ precharge reduces the read time. However, precharging

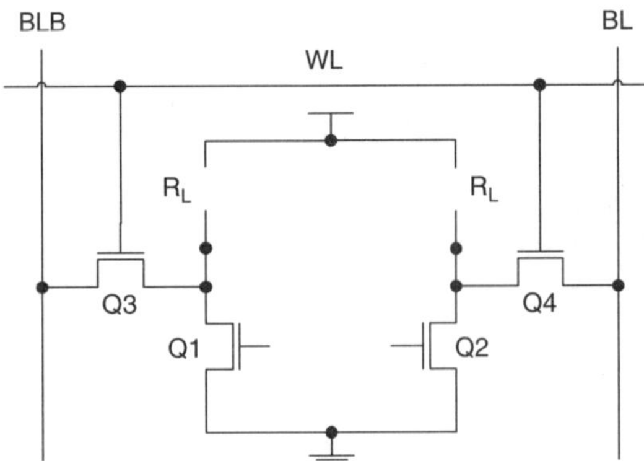

Fig. 2.2 Four-transistor (4T) SRAM cell with polysilicon resistor load

the bit lines to $V_{DD}/2$ can compromise the cell stability. Precharge of bit lines to full V_{DD} can alleviate the requirement for the low-to-high cell transition current at the cost of the additional precharge time and the associated power consumption. The upper resistance limit on R_L is set by the requirement to provide a pull-up current of at least two orders of magnitude larger than the leakage current of the pull-down transistor [21]. The lower limit on R_L is set by the required noise immunity and power consumption requirements. The technological variations of R_L caused by the limitations of doping and annealing techniques pose another constraint on the increase of R_L.

As the technology scaled into sub-micron regime (beyond 0.8 μm technology generation), the scalability of a PRL SRAM cell became an issue [22]. The polysilicon resistor in the PRL cell could not be scaled as aggressively as the cell's transistors. The switch from constant-voltage scaling to constant-field scaling to combat the short-channel effects and electric field implications on the long-term reliability, revealed non-satisfactory low-voltage power consumption vs. stability performance tradeoff of the PRL cells. Moreover, the extra technological steps of forming high-resistivity polysilicon are not a part of the standard CMOS logic technological process. Insufficient tolerance to soft errors, which can be linked to an inadequate SNM, adds to the list of disadvantages of a PRL cell. These factors prohibit using the PRL SRAM cells in Systems-on-a-Chip (SoCs) that are traditionally implemented using standard full CMOS processes. All the mentioned factors practically excluded the PRL cell from being used the current mainstream scaled-down deep-submicron technologies.

2.3.2 Six-Transistor (6T) CMOS SRAM Cell

The mainstream six-transistor (6T) CMOS SRAM cell is shown in Figure 2.3. Similarly to one of the implementations of an SR latch, it consists of six transistors. Four transistors ($Q1-Q4$) comprise cross-coupled CMOS inverters and two

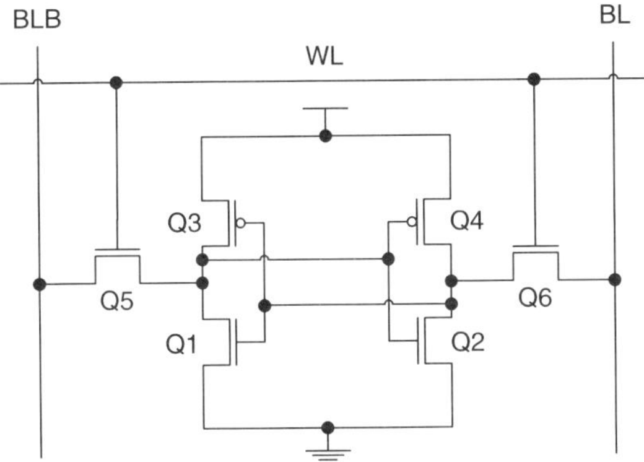

Fig. 2.3 Six-transistor (6T) CMOS SRAM cell

NMOS transistors $Q5$ and $Q6$ provide read and write access to the cell. Upon the activation of the word line, the access transistors connect the two internal nodes of the cell to the true (BL) and the complementary (BLB) bit lines.

A 6T CMOS SRAM cell is the most popular SRAM cell due to its superior robustness, low power and low-voltage operation. Therefore, we will discuss its operation and design in greater detail. An SRAM cell must be designed such that it provides a non-destructive read operation and a reliable write operation. These two requirements impose contradicting requirements on SRAM cell transistor sizing. SRAM cell transistor ratios that must be observed for successful read and write operations are discussed in the following sections.

2.3.2.1 Read Operation

Prior to initiating a read operation, the bit lines are precharged to V_{DD}. The read operation is initiated by enabling the word line (WL) and connecting the precharged bit lines, BL and BLB, to the internal nodes of the cell.

Upon read access shown in Figure 2.4, the bit line voltage V_{BL} remains at the precharge level. The complementary bit line voltage V_{BLB} is discharged through transistors $Q1$ and $Q5$ connected in series. Effectively, transistors $Q1$ and $Q5$ form a voltage divider whose output is now no longer at zero volt and is connected to the input of inverter $Q2-Q4$ (Figure 2.3). Sizing of $Q1$ and $Q5$ should ensure that inverter $Q2-Q4$ does not switch causing a destructive read. In other words, $0+\Delta V$ should be less than the switching threshold of inverter $Q2-Q4$ plus some *safety margin* or *Noise Margin*.

Ignoring the short-channel and body effects, the maximum allowed value $0+\Delta V$ of the node storing a logic "0" during read access can be expressed as [21]:

$$\Delta V = \frac{V_{DSATn}+CR(V_{DD}-V_{THn})-\sqrt{V_{DSATn}^2(1+CR)+CR^2(V_{DD}-V_{THn})^2}}{CR} \qquad (2.1)$$

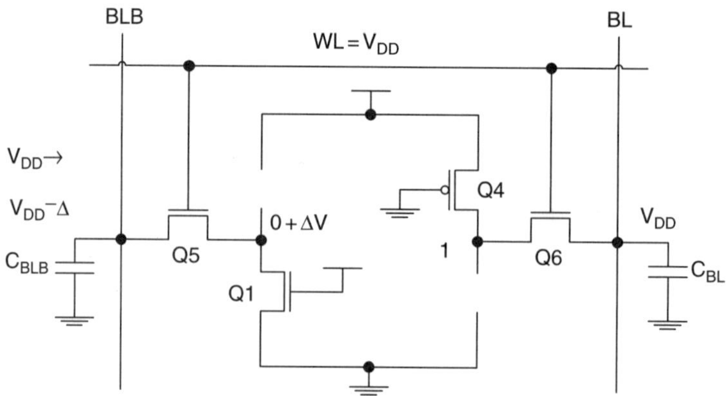

Fig. 2.4 Simplified model of a 6T CMOS SRAM cell during a read operation

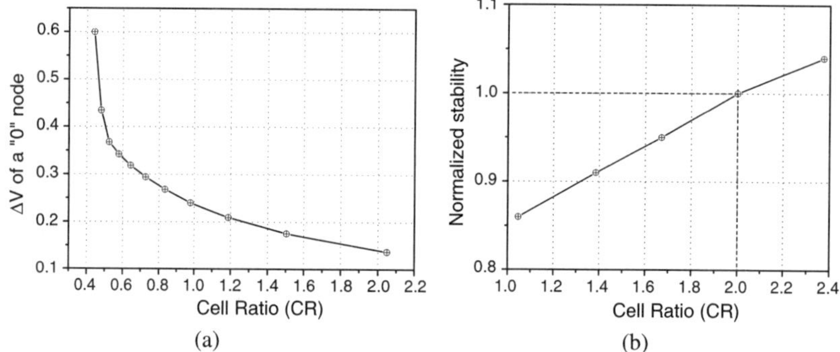

Fig. 2.5 The rise ΔV of the "0" node (a) and the SNM (b) as a function of the Cell Ratio (CR) $\left(CR = \frac{W_1}{L_1} / \frac{W_5}{L_5} = \frac{W_2}{L_2} / \frac{W_6}{L_6} \text{ in Figure 2.3}\right)$ in a 6T CMOS SRAM cell (simulated in CMOS 0.13 µm technology, $V_{DD} = 1.2\,\text{V}$)

where CR (β) is the cell ratio defined as

$$\beta = CR = \frac{W_1/L_1}{W_5/L_5} \tag{2.2}$$

Since the cell is fully symmetrical, the CR is the same for $Q2$ and $Q6$.

The dependence of ΔV on the CR is shown in Figure 2.5(a). Typically, in order to ensure a non-destructive read and an adequate noise margin, CR must be greater than one and can be varied depending on the target application of the cell from approximately 1 to 2.5. Larger CRs provide higher read current I_{read} (and hence the speed) and improved stability (Figure 2.5(b)) at the expense of larger cell area. A smaller CRs ensure a more compact cell with moderate speed and stability. Leakage through the access transistors should be minimized to ensure robust read operation and to reduce the leakage power.

A preferred sizing solution can be to use a minimum-width access transistors with a slightly larger than the minimal length channel and a larger than minimal width with a minimal length driver transistors.

Once the complementary bit line discharges to a certain $V_{BLB} = V_{DD} - \Delta V$ voltage level sufficient for reliable sensing by the sense amplifier, the sense amplifier is enabled and amplifies the small differential voltage between the bit lines to the full-swing CMOS level output signal.

2.3.2.2 Write Operation

The write operation is similar to a reset operation of an SR latch. One of the bit lines, BL in Figure 2.6, is driven from precharged value (V_{DD}) to the ground potential by a write driver through transistor $Q6$. If transistors $Q4$ and $Q6$ are properly sized, then the cell is flipped and its data is effectively overwritten. A statistical measure of SRAM cell writeability is defined as *write margin*. Write margin is defined as the minimum bit line voltage required to flip the state of an SRAM cell [23]. The write margin value and variation is a function of the cell design, SRAM array size and process variation. A cell is considered not writeable if the worst-case write margin becomes lower than the ground potential.

Note that the write operation is applied to the node storing a "1". This is necessitated by the non-destructive read constraint that ensures that a "0" node does not exceed the switching threshold of inverter $Q2 - Q4$. The function of the pull-up transistors is only to maintain the high level on the "1" storage node and prevent its discharge by the off-state leakage current of the driver transistor during data retention and to provide the low-to-high transition during overwriting.

Assuming that the switching will not start before "1" node is below V_{TH_Q1}, a simplified overwrite condition can be expressed as [21]:

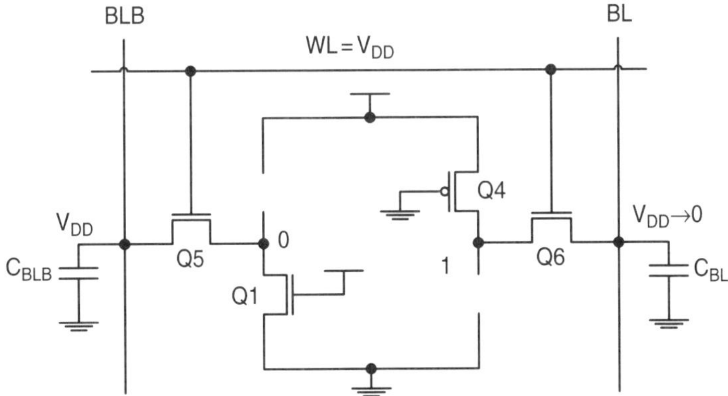

Fig. 2.6 Simplified model of a 6T CMOS SRAM cell during a write operation

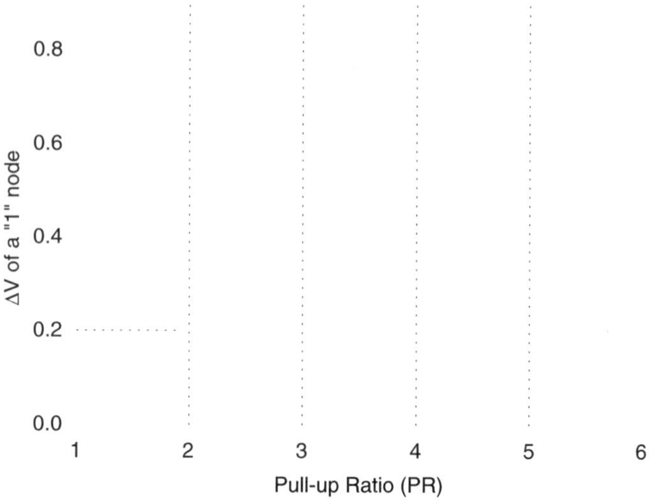

Fig. 2.7 The voltage drop at node $V_{“1”}$ during write access as a function of the Pull-Up ratio (PR) $\left(PR = \frac{W_4}{L_4} / \frac{W_6}{L_6} = \frac{W_5}{L_5} / \frac{W_3}{L_3} \text{ in Figure 2.3}\right)$ of a 6T CMOS SRAM cell (simulated in CMOS 0.13 μm technology, V_{DD} = 1.2 V)

$$V_{“1”} = V_{DD} - V_{THn} - \sqrt{(V_{DD} - V_{THn})^2 - 2\frac{\mu_p}{\mu_n} PR\left((V_{DD} - |V_{THp}|)\,V_{DSATp} - \frac{V_{DSATp}^2}{2}\right)} \tag{2.3}$$

where the pull-up ratio of the cell, PR, is defined as:

$$PR = \frac{W_4/L_4}{W_6/L_6} \tag{2.4}$$

The $V_{“1”}$ requirement is normally met using minimal-sized access and pull-up transistors only due to μ_n/μ_p ratio. Simulation results shown in Figure 2.7 demonstrate that for a normal write operation, i.e., to pull the $V_{“1”}$ node below V_{THn}, the W/L of the pull-up transistor must be less than 3-4 W/L of the access transistor. The exact maximum allowed PR is defined by the V_{THn} process option and by the switching threshold of inverter $Q1 - Q3$ in Figure 2.3.

Normally, to minimize the cell area and hence, increase the packing density, the sizes of the pull-up and access transistors are chosen to be minimal and approximately the same. However, stronger access transistors and/or weaker pull-up transistors may be needed to ensure a robust write operation under the worst process conditions e.g., in the fast PMOS and slow NMOS process skew corner. On the other hand, a relatively strong pull-up PMOS also benefits the read stability due to the increased P/N ratio of the back-to-back inverters ($Q3 - Q1$ and $Q4 - Q2$ in Figure 2.3) of the cell's latch. The read stability of an SRAM cell on one hand and the writeability of the cell on the other hand are conflicting design requirements. It is getting increasingly more difficult to balance these requirements by conventional transistor sizing and V_{TH} optimization as the design window becomes increasingly narrower [23] with the technology scaling.

Despite the larger number of transistors compared to the other discussed cells, 6T CMOS SRAM cells offer superior stability and packing density provided the same performance and environmental tolerance.

2.3.3 Four-Transistor (4T) Loadless SRAM Cell

Recently, the new loadless 4T CMOS SRAM cell shown in Figure 2.8(b) was proposed by NEC [24, 25] for ultra-high density SRAM macros [26, 27]. A 4T SRAM cell uses minimal-size NMOS (Q1, Q2) and PMOS (Q3, Q4) transistors. Data retention without a refresh operation is ensured if the leakage current of the PMOS transistors is higher than the leakage current of the NMOS transistors. This condition is normally met by using dual-V_{TH} process with $V_{TH\_p} < V_{TH\_n}$. For stable data retention and to account for the I_{off} process distribution, the leakage current of the PMOS transistors in a 4T loadless cell must be around 100 times higher than that of the driver NMOS transistors. In a 4T SRAM cell, PMOS transistors Q3 and Q4 serve as access transistors as opposed to NMOS access transistors Q3 and Q4 in a 6T SRAM cell (Figure 2.8(a)). Due to the mobility ratio of the NMOS to the PMOS transistors μ_n/μ_p which is normally around two to three, all the transistors in a 4T SRAM cell can be of minimal size. Recall that for a 6T SRAM cell to guarantee a non-destructive read operation, NMOS driver transistors Q1 and Q2 must be 1.5–2.5 times larger than NMOS access transistors Q3 and Q4 in Figure 2.8. This fact, in addition to the larger number of transistors in a 6T SRAM cell, makes a 4T SRAM cell an area-efficient choice.

For the same access speed and comparable SNM, the area of a 4T SRAM cell is 50–65% of the area of a conventional 6T SRAM cell. Since memory blocks occupy

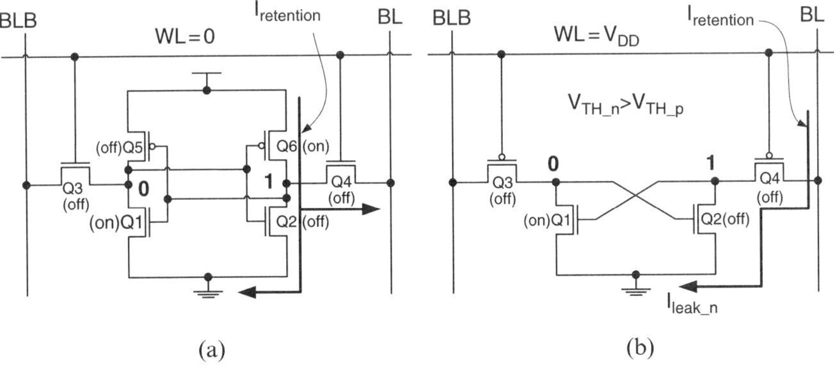

Fig. 2.8 Six-transistor (a) and four-transistor (b) CMOS SRAM cells

considerable chip area in SoCs, SRAM cell area is a critical factor in the SoC design. The area savings offered by the 4T SRAM cells have been one of the main driving forces behind the 4T CMOS SRAM cell development.

However, 4T CMOS SRAM cells are not without drawbacks. Reliable data retention in a 4T cell can be guaranteed only if I_{leak_p} is significantly larger than I_{leak_n} under the worst case PVT conditions. This condition is increasingly more difficult to fulfill in nano-scaled processes with higher process parameter variations. The size advantage of 4T CMOS SRAM cells may be compromised by the stability degradation caused by the growing statistical spread of transistor parameters in nano-scaled technologies. Another disadvantage of 4T CMOS SRAM cells can reveal itself during a write operation. When writing to a certain column, the write driver drives the bit line from V_{DD} to $0V$. During this time, PMOS transistors in the other cells in the same column cannot provide the leakage current I_{PMOS_leak} to maintain the nodes storing "1". If the ratio of precharge time to the cycle time is not sufficient and the same memory address is being successively write accessed, there is a risk that other cells in the same column may flip.

2.4 Cell Layout Considerations

The layout of an SRAM cell defines the area density of the array and is key to manufacturing yield of the SoC containing large SRAM arrays. As the process technology continues to scale down, the limitations of lithography, etching and CMP call for modifications in the layout of SRAM cells. The general trend in the cell layout modifications is to streamline the patterns. By avoiding complex geometries, the reproducibility of the minimum feature sized patterns that define W and L of SRAM cell transistors can be greatly improved.

Four-transistor (4T) SRAM cells with a resistive load have dominated the stand-alone SRAMs since first introduced in the 1970s to around mid-1990s [28]. The stand-alone SRAM that used 4T cells often employed a special process, which allowed to shrink the cell size at the price of more process steps. However, the process for stand-alone SRAM often was not compatible with the logic process. The 6T SRAM cells, which are extensively used for on-die caches, are fabricated using the same process steps which are used for the logic, which warranted their use as embedded memory for microprocessors. The larger cell area due to two extra transistors was the main factor that initially prevented wide use of the 6T cell in the stand-alone applications.

A schematic and a sample layout of a 4T cell with a resistive load is shown in Figure 2.9(a). The cell uses NMOS driver ($Q1$ and $Q2$) and access ($Q3$ and $Q4$) transistors. The state of the "1" data node was preserved by a highly-resistive undoped polysilicon load resistors laid out on top of the cell (not shown in the layout). The pull-up current provided by the polysilicon resistors was chosen as a trade-off between the cell stability and the dissipated power per cell. If the resistance of the load is too large, the leakage current of the driver transistor can discharge the

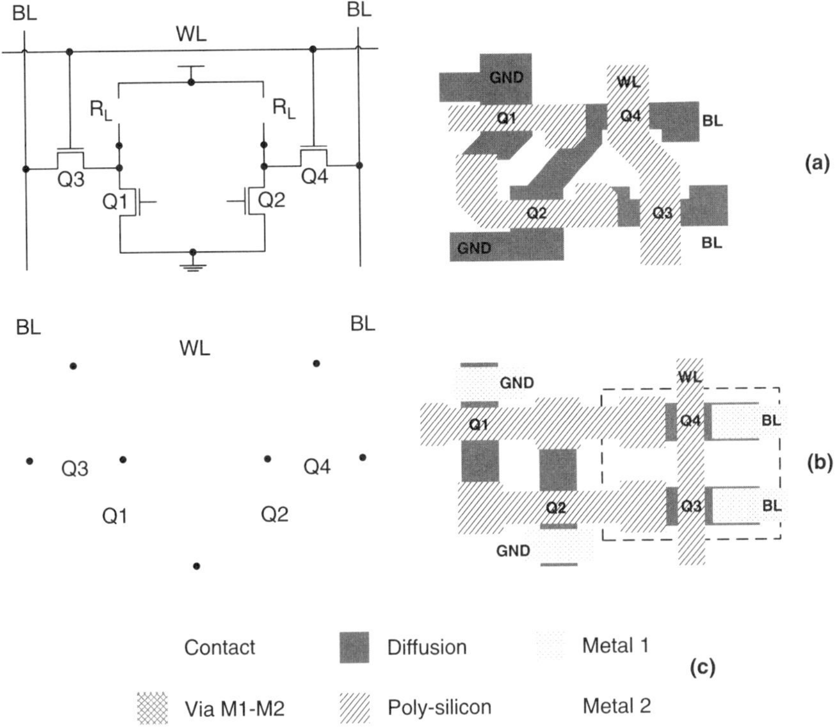

Fig. 2.9 Schematics and sample layouts of 4T SRAM cells: 4T polysilicon resistor load cell (a); 4T loadless cell (b); layout layers legend (c)

"1" data node. On the other hand, if the resistance is too small, the resulting voltage on the "1" data node caused by the voltage divider action of the load resistor and the driver transistor could partially turn the driver transistor of the other cell inverter on. Moreover, the power dissipation per cell in case of a small load resistor value multiplied by the number of cells in the array can become a significant part of the total power dissipation.

In an attempt to improve the insufficient stability margin in technology generations with the power supply voltage below 5 V, the 4T SRAM cells with polysilicon resistor load are replaced by the cells with TFT-transistor load. The "on" resistance of TFT transistors is lower than the "off" resistance, helping to reduce the cell size and the power and improve the noise immunity.

The schematic and a sample layout of another reincarnation of a 4T cell – a 4T loadless cell – is shown in Figure 2.9(b). The operation of this cell is described in more details in Section 2.3.3. The functions of the load and access transistors of this 4T cell are performed by two PMOS transistors (*Q*3 and *Q*4). SRAM designs using these cells have been reported for 180 nm [27, 29] and 130 nm [26] process technologies. Since the cell relies on the ratio of $\frac{I_{off}PMOS}{I_{off}NMOS}$, the data retention capabilities of this cell is strongly dependent on a strict process control.

By the mid-1990s and process technology generations with a feature size below 0.35 μm [28], the industry had mostly migrated from 4T cells with a resistive or TFT load to 6T CMOS cells for both the stand-alone and the on-chip storage. The constant electric field technology scaling drove the supply voltages below 5 V and became the main reason of the change in the cell type and layout. The 6T CMOS SRAM cells demonstrated the best stability in low-voltage operation while the advances in the process technology helped to improve the area efficiency of 6T cells.

A schematic and layout types of a 6T CMOS SRAM cells are shown in Figure 2.10. With a PMOS transistor as an active load, the stability margin and the power dissipation of an SRAM cell dramatically improved. Due to its form factor, the layout in Figure 2.10(a) is coined "tall". Tall layouts of SRAM cells were widely used down to the 90 nm technology generation. Due to high capital costs, the lithography equipment was updated at a slower pace than the progress of the transistor scaling, which followed the Moore's Law. Reproducible pattering of the design features that are much smaller than the wavelength of the lithography light source became problematic. The growing number of cells in the scaled SRAM arrays together with poor pattern reproducibility led to a higher number of marginal cells in the array which can reduce quality and yield and the efficiency of the power reduction techniques. Further scaling of a "tall" cell became more problematic due to poor reproducibility of the multiple pattern corners in the layout. Therefore, beyond the 90 nm technology generation node, the "tall" SRAM cell

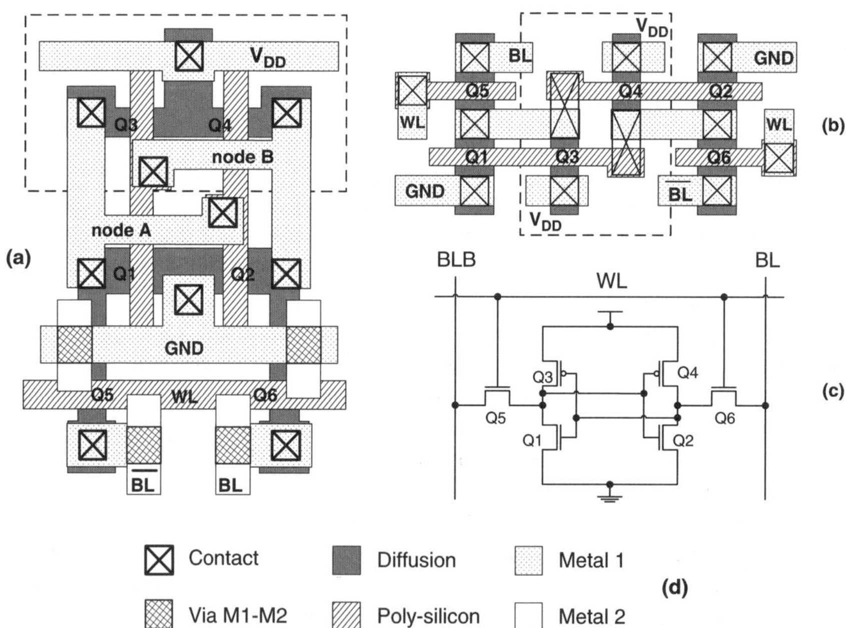

Fig. 2.10 Schematic and sample layouts of 6T SRAM cells: 6T CMOS "tall" cell (a); 6T CMOS "wide" cell (b); schematic of a 6T CMOS SRAM cell (c); layout layers legend (d)

layout shown in Figure 2.10(a) has been commonly replaced by the "wide" layout shown in Figure 2.10(b) [30–33]. The much streamlined layout of a "wide" cell and the uniform orientation of all the transistors in the cell provide better pattern reproducibility (W and L matching) and transistor V_{TH} matching. The impact and importance of good transistor matching in an SRAM cell is described in more detail in Chapter 3. Since the word lines in the "wide" bit layout are implemented in metal, the "wide" cell layout can offer reduced word line resistance and shorter propagation time of the WL signal to the parts of SRAM array which are further from the word line drivers. Another benefit of migrating to the "wide" SRAM cell is the shorter bit line length per cell. The shorter bit line length per cell transfers to the lower bit line capacitance C_{BL}. In turn, a lower C_{BL}/bit can be beneficial in one of the following ways. Firstly, it can be transferred to a faster developing of the bit-line differential signal and, thus, helps to achieve a faster cycle time of an array built of "wide" SRAM cells. Secondly, it increases the number of cells per bit line while keeping the same bit-line differential signal slew rate and, thus, helps to improve the array efficiency.

Diffusion, poly, contact and metal 1 are the four masking layers considered critical in the processing of SRAM cells. The tightened design rules applied to the SRAM array in these layers increase the risk of bridging/shorting. The fidelity of the patterns in these layers is, however, critical to cell stability, drive and leakage currents. To improve the pattern fidelity, the patterns in deep-submicron processes are corrected using Optical Proximity Correction (OPC). OPC of SRAM cells is especially challenging considering the tighter design rules. SRAM OPC often apply the accumulated knowledge of lithographic conditions and properties of the photoresist to be employed for each masking layer and a careful study and simulation of the resulting photoresist pattern on Si. The automated OPC algorithms though more generic, help to account for the process condition modifications [34]. Table 2.1 summarizes the issues associated with the critical layers in SRAM cell layout.

Table 2.1 Layout issues addressed during SRAM cell design

Layer	Issues addressed
Diffusion	(a) Critical Dimensions (CDs) (for I_{read}) (b) Bridging (c) Adequate area for contact (for low $R_{contact}$)
Poly	(a) Critical Dimensions (CDs) (for I_{read}) (b) Symmetry (for SNM) (c) Bridging (d) Adequate end-cap
Contact	(a) Adequate coupled contact coverage over diffusion and poly (b) Bridging with adjacent contacts (c) Adequate clearance from potential ILD voids
Metal 1	(a) Bridging (b) Adequate overlap of contact and via (for low $R_{contact}$)

Depending on the target application, SRAM cells may undergo layout modifications. While optimized for maximum bit density and minimum cell leakage, the cells for high density (HD) applications may have smaller area than cells for high speed (HS) applications. HS cells need larger driver transistors to provide larger cell currents and improved SNM.

2.5 Sense Amplifier and Bit Line Precharge-Equalization

Sense amplifiers (SA) are an important component in memory design. The choice and design of a SA defines the robustness of bit line sensing, impacting the read speed and power. Due to the variety of SAs in semiconductor memories and the impact they have on the final specs of the memory, the sense amplifiers have become a separate class of circuits.

The primary function of a SA in SRAMs is to amplify a small analog differential voltage developed on the bit lines by a read-accessed cell to the full swing digital output signal thus greatly reducing the time required for a read operation. Since SRAMs do not feature data refresh after sensing, the sensing operation must be nondestructive, as opposed to the destructive sensing of a DRAM cell. A SA allows the storage cells to be small, since each individual cell need not fully discharge the bit line.

Design constraints for an SA are defined by the minimum differential input signal amplitude, the minimum gain, A, and tolerance to the environmental conditions and mismatches. The gain is a function of the initial bit line voltage (precharge level). In addition, it influences the sense delay t_{sense}. However, a high A does not necessarily reduce t_{sense}. Usually, t_{sense} is traded off for reduced power consumption, layout area and for improved tolerance of environmental conditions [17].

Special attention is given to the SA area. Depending on the SRAM array column architecture, the area requirements for the SA may vary. Architectures using column multiplexing can share a single SA among the multiplexed columns such that only one column is connected to the sense amplifier at any given time. The total area available for the SA is defined by a multiple n of the bit line pitch values, where n can normally be from 1 to 16. In turn, the bit line pitch is defined by the size of a memory cell. This example illustrates the complexity of SRAM design and layout planning. The choice of the cell size, number of columns, number of cells per column, the minimum differential swing, the choice of the SA architecture and size are all the factors taken into consideration when designing an SRAM compliant with the target power, speed and reliability.

Generally, the parameters characterizing a sense amplifier include:

- Gain $A = V_{out}/V_{in}$
- Sensitivity $S = V_{in\_min}$ – minimum detectable signal
- Offsets V_{offset} and I_{offset} – the difference at the outputs with the common mode signal at the inputs

- Common Mode Rejection Ratio $CMRR = A_{diff}/A_{cm}$ – ratio of amplification for a differential and a common mode signals
- Rise time t_{rise}, fall time t_{fall} – 10% to 90% of the signal transient
- Sense delay $t_{sense} = t_{50\%\_WL} - t_{50\%\_V\_out}$ – where $t_{50\%\_WL}$ – the 50% point of the word line enable signal and $t_{50\%\_V\_out}$ – the 50% point of SA output transient

The simultaneous optimization of the above parameters is a difficult task involving balancing the circuit complexity, layout area, reliability, power, speed and environmental tolerance. The process spread of the modern DSM technologies can add to the complexity of the SA design by introducing significant parameter mismatches, asymmetry and offsets. Practical SA design is an iterative procedure that must pay close attention to the fabrication process parameters and their variations in the target technology. The choice of circuit, transistor sizing, operating point, gain and transient response must be done based on the timing and layout constraints of the particular memory system. To alleviate short-channel effects and the atomic dopant distribution effects in the channel [35], SAs often employ devices with non-minimum length and width. That helps to reduce the asymmetry resulting from transistor V_{TH} and geometry spread and thus mitigate the SA offset caused by the inherent variation in the parameters of the fabricated transistors.

The bit line differential voltage, the reliability and the power consumption of an SA are directly linked. The minimal bit line differential is a factor in defining the total read access time and thus, the speed of an SRAM. On one hand, a larger bit line differential is beneficial for more reliable sensing. On the other hand, the resulting better tolerance to the process and environmental fluctuations comes at the cost of the extra read access time and the power spent on the discharging and precharging of the bit lines.

The differential sensing, widely used in SRAMs, allows rejection of the common-mode noise that may present on both the bit lines. Noise sources, such as power spikes, capacitive coupling between the bit lines and between the word line and the bit lines, can inject common-mode noise to both SA inputs. The common mode noise is then attenuated by the value of ratio CMRR and the true differential signal is amplified.

A classical current-mirror differential SA with active load is shown in Figure 2.11. The sensing operation begins with setting the SA operation point by precharging and equalization of both the inputs of the SA (which are the bit lines *BL* and *BLB* in Figures 2.11, 2.12) to the identical *precharge* voltage level. Once both *BL* and *BLB* are precharged and equalized, the precharged levels are stored in the bit line capacitance C_{BL}. Next, the decoded word line WL of a read-accessed cell is activated starting the build-up of the differential voltage on the bit lines *BL* and *BLB* (around 100–200 mV). Once the differential voltage exceeds the sensitivity S of the SA, a Sense Amplifier Enable (SAE) signal is issued and the SA amplifies the differential voltage on the bit lines to the full-swing output level *out*. Then, the *SAE* and WL are deasserted and the read operation is complete.

For reliable operation, current mirror SAs require biasing power to set up in the high-gain region. The minimum biasing current is limited by the minimum required SA gain, CMRR and sensing delay, whereas the maximum biasing current is

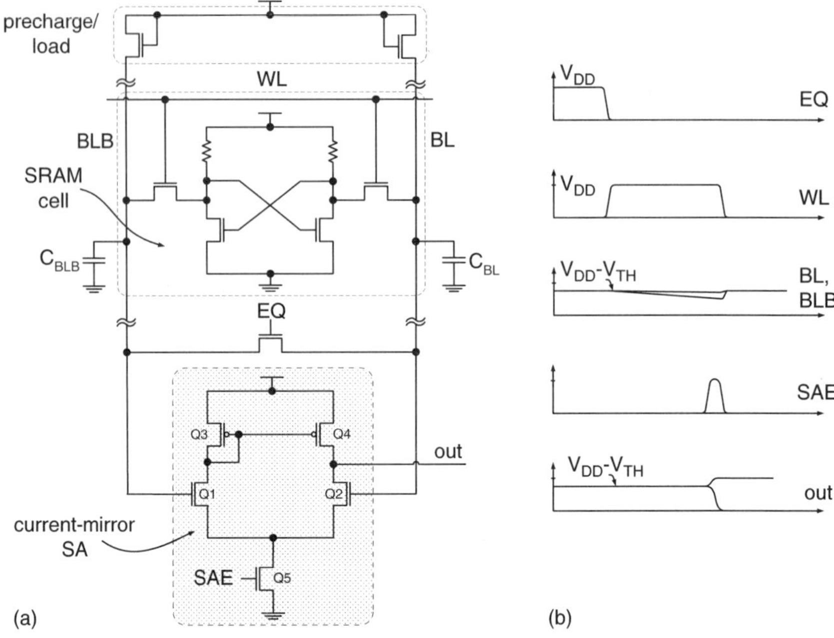

Fig. 2.11 A typical circuit with a current-mirror type sense amplifier, a PRL SRAM cell and precharge/load transistors (a); signal waveforms during a read operation (b)

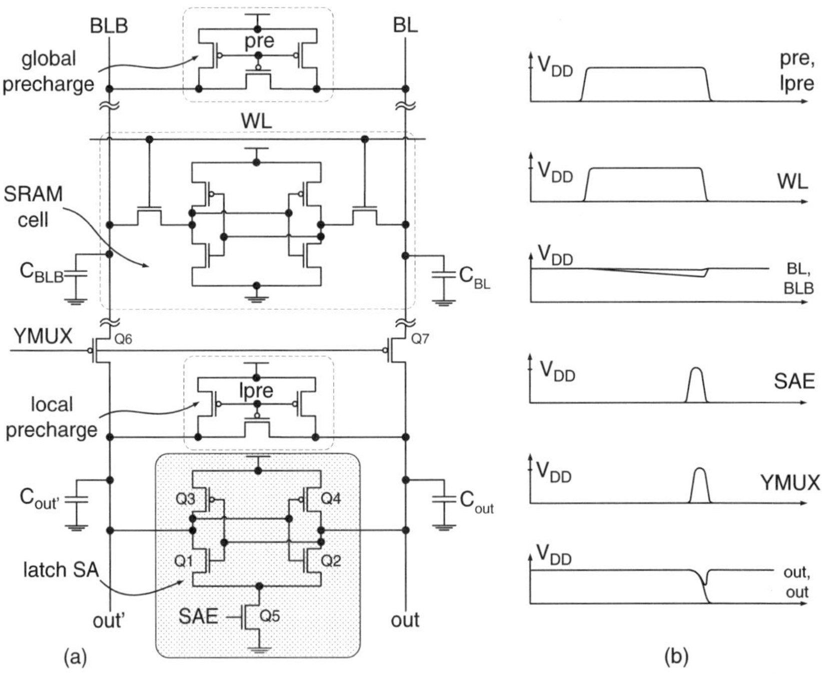

Fig. 2.12 A typical circuit with a latch-type sense amplifier, a full CMOS 6T SRAM cell, column mux and precharge (a); signal waveforms during a read operation (b)

limited by the power dissipation and the layout area. The gain of a current-mirror SA is given by Equation 2.5. The gain is typically set around ten. The gain can be increased by widening $Q1$ and $Q2$ or by increasing the biasing current [21]:

$$A = -g_{m\_Q1}(r_{o2}||r_{o4}) \tag{2.5}$$

where g_{m_Q1} is the transconductance of $Q1$, and r_{o2} and r_{o4} are small-signal output resistance of $Q2$ and $Q4$, respectively.

The output resistance of the current-mirror SA is given by $R_{out} = r_{o2}||r_{o4}$. Parameters of g_m and r can be modified by the proper transistor sizing. To ensure high initial amplification, the sizing of $Q1 - Q4$ is chosen such that the bit line precharge level corresponds to the high-gain region of the SA's transfer characteristic. However, the gain A is also a function of the V_{GS} of transistors $Q1$ and $Q2$, and hence a function of the precharge level. The precharge NMOS transistors used to statically precharge the bit lines should be sized so that their contention with the pull-down transistors does not flip the cell. That puts a sizing constraint on the precharge/load transistors. The sizing of these transistors determines the bit line recovery speed, which is especially critical after a write operation when the bit line is completely discharged. High sensitivity to transistor mismatches in a current-mirror SA causes increased offsets. To compensate for possible offsets and maintain reliable sensing, the minimum differential voltage must be increased, slowing down the sensing. Combined with the sizable power consumption and special precharge conditions, this causes the usage of the current-mirror type SA to decline in the scaled-down low-voltage technologies. The circuit in Figure 2.11 does not require SA isolation as the bit lines are connected to the transistor gates and are isolated from the output. The voltage-divider action of the serially-connected driver, access and the precharge/load transistors prevents the complete discharge of the bit lines. Thus, the word line deactivation timing requirements can be relaxed as the bit line discharge will stop at the potential defined by the relative sizing of the precharge/load, access and driver transistors.

A latch-type SA is shown in Figure 2.12. This type of a SA is formed by a pair of cross-coupled inverters, much like a 6T SRAM cell. The sensing starts with biasing the latch-type SA in the high-gain metastable region by precharging and equalizing its inputs. Since in the latch-type SA the inputs are not isolated from the outputs, transistors $Q5$ and $Q6$ are needed to isolate the latch-type SA from the bit lines and prevent the full discharge of the bit line carrying a "0", which costs extra power and delay. Due to the presence of the column MUX/isolation transistors, two precharge/equalize circuits are needed to ensure reliable sensing: global precharge/equalization *pre* for the column and a local precharge/equalization *lpre* for the inputs of the SA (Figure 2.12 (a)).

When a cell accessed by the word line WL has discharged the bit lines BL and BLB to a sufficient voltage differential (Figure 2.12 (b)), the SA is enabled by a high-to-low transition of SAE pulse. Shortly after that, the column MUX/isolation transistors $Q6$ and $Q7$ are turned off, isolating the highly capacitive bit lines from the SA latch and preventing the complete discharge of C_{BL} and C_{BLB}. Then, the positive feedback of the cross-coupled inverters $Q1 - Q3$ and $Q2 - Q4$ quickly drives the low-capacitance outputs *out* and out' to the full swing complementary voltages. Note that in the typical circuit presented in Figure 2.12, the local and global precharge/equalize circuits are clocked to save power. To improve the noise robustness in low-voltage operation and to ensure a non-destructive read under process variations and minor defects in the cell in modern DSM technologies, the precharge level is typically set to full V_{DD}.

2.6 Write Driver

The function of the SRAM write driver is to quickly discharge one of the bit lines from the precharge level to below the write margin of the SRAM cell. Normally, the write driver is enabled by the Write Enable (WE) signal and drives the bit line using full-swing discharge from the precharge level to ground. The order in which the word line is enabled and the write drivers are activated is not crucial for the correct write operation.

Some of the typical write driver circuits are presented in Figure 2.13. The circuit in Figure 2.13(a) writes the input data *in* and its complement buffered by inverters 2 and 3 to the bit lines *BL* and *BLB* through two transmission gates $TG1$ and $TG2$. WE and its complementary WEB are used to activate $TG1$ and $TG2$ and discharge *BL* or *BLB* through the NMOS transistors in inverter 2 or 3. The write driver presented in Figure 2.13(b) uses two stacked NMOS transistors to form two pass-transistor AND gates using transistors $Q1, Q3$ and $Q2, Q4$. The sources of NMOS transistors $Q1$ and $Q2$ are grounded. When enabled by WE, the input data *in* enables, through inverters 1 and 2, one of the transistors $Q1$ or $Q2$ and a strong "0" is applied by discharging *BL* or *BLB* from the precharge level to the ground level. Another implementation of the write driver is presented in Figure 2.13(c). When WE is asserted, depending on the input data *in*, inverters 1 and 2 activate one of two two-input AND gates 1 and 2 to turn on one of the pass-transistors $Q1$ or $Q2$. Then, the activated transistor discharges the corresponding bit line to the ground level.

Even though a greater discharge of the highly capacitive bit lines is required for a write operation, a write operation can be carried out faster than a read operation.

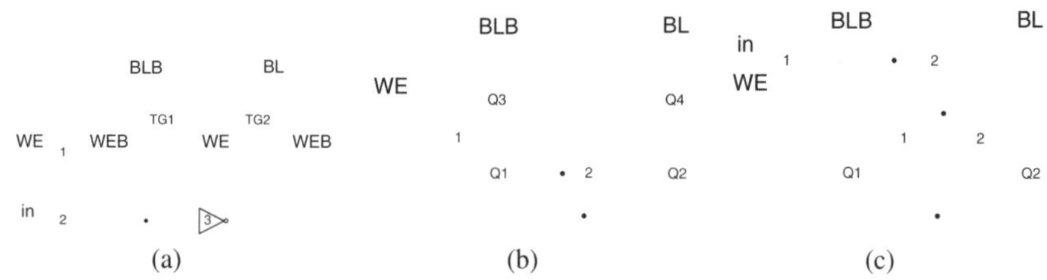

Fig. 2.13 Write driver circuits

Only one write driver is needed for each SRAM column. Thus, the area impact of a larger write driver is not multiplied by the number of cells in the column and hence the write driver can be sized up if necessary.

2.7 Row Address Decoder and Column MUX

Address decoders allow the number of interconnects in a binary system to be reduced by a factor of $\log_2 N$, where N is the number of independent addressed locations. The memory address space is defined as the total number of the address bits required to access a particular memory cell or word for bit-oriented memories and word-oriented memories, respectively. For instance, the total address space in a 1Mb bit-oriented SRAM will be 20 (2^{20} = 1 Mb) address bits A0...A19. On the other hand, in a 1Mb word-oriented SRAM with a 32-bit (2^5) word width, which can be organized in 32 blocks each of which has 256 rows and 128 columns, the address space reduces to 15 ($2^{(20-5)} = 2^{15}$) address bits A0...A14.

The SRAM row decoder can be of a single- or multi-stage architecture. In a single-stage decoder all decoding is realized in a single block. The multiple-stage decoding uses several hierarchically-linked blocks. Normally, the most significant bits are decoded ($pre-decoded$) in the first decoder stage, effectively selecting the array that is to be accessed by providing enable signals for the subsequent decoder stage(s) that enable a particular word line. The number of outputs of the last decoding stage corresponds to the number of rows (word lines) to be decoded. An example of single-stage static decoders is shown in Figure 2.14(a). Each word line is decoded in a single stage by a wide NOR gate with a fan-in equal to the number of row address bits. To simplify the circuit and reduce the layout area, such decoders are often designed using a static PMOS transistor load (top circuit in Figure 2.14(a)). Another variation of this decoder uses a clocked precharge PMOS transistor (bottom circuit in Figure 2.14(a)).

Single-stage row decoders are attractive for small single-block memories. However, most memories today split the row address space into several blocks decoded by separate decoder stages. This approach is proven to be more power efficient and faster for large memories with of multiple arrays. An example of a multi-stage 4–16 decoder shown in Figure 2.14(b) uses two 2–4 decoders and a 8–16 stage. Address bits A0, A1 and A2, A3 are predecoded separately by the 2–4 decoders, which drive eight address lines. A set of 16 AND gates further decodes the predecoded combinations of A0, A1 and A2, A3 into 16 outputs.

All large decoders use at least a two-stage implementation [21]. The conventional Divided Word Line (DWL) structure shown in Figure 2.15(a) partitions the SRAM into blocks. A local (block) word line is activated when both the global word line and the block select are asserted. Since only one block is activated, the DWL structure reduces both the word line delay and the power consumption. An additional decoding level, coined Hierarchical Word Decoding (HWD) architecture, was proposed for larger than 4 Mb SRAMs (Figure 2.15(b)) to cope with the growing delay

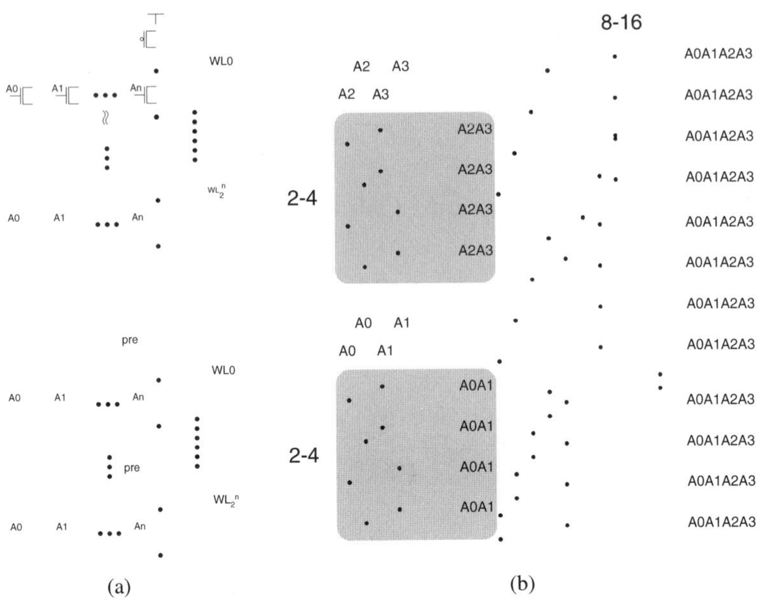

Fig. 2.14 (a) Single-stage static (top) and dynamic (bottom) decoders; (b) Multi-stage static 4-16 decoder

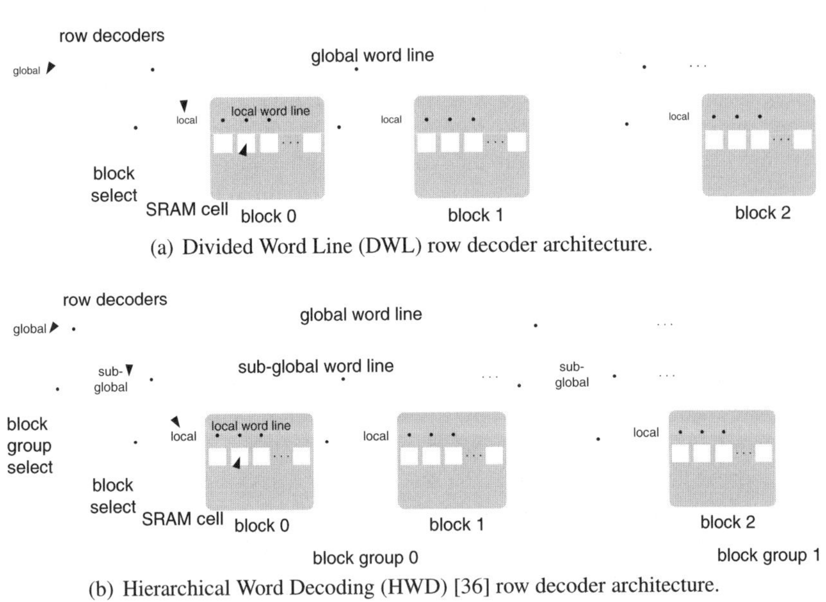

Fig. 2.15 Multi-stage row decoder architectures

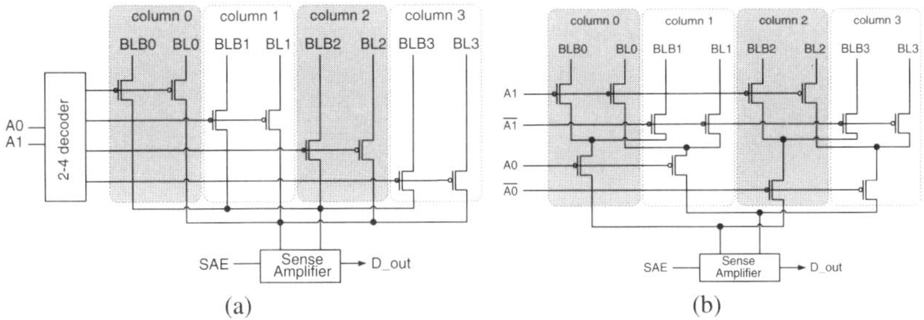

Fig. 2.16 (a) 4-1 pass-transistor column decoder with a predecoder; (b) 4-1 tree-based column decoder

time and power consumption [36]. The HWD offers $\approx 20\%$ delay and $\approx 30\%$ total load capacitance reduction over the DWL decoder architecture.

A column MUX in an SRAM uses a 2^K-input multiplexer for each of the bit lines, where K is the size of column address word. A column MUX allows several columns to be connected to a single SA and thus, relax the area constraints on the SA design. An example of a typical column MUX using pass-transistors and a 2–4 predecoder is shown in Figure 2.16(a). PMOS transistors enabled by one of the outputs of the 2–4 predecoder pass the read differential voltage from the bit lines of one out of four columns to the inputs of a differential SA. The simpler version of a column MUX shown in Figure 2.16(b) uses a binary-tree decoder of PMOS pass transistors. This column MUX requires no predecoding and uses fewer transistors. However, since the propagation delay of the column MUX increases quadratically with the number of sections, a large tree-based column MUX introduces extra delay and that may bc prohibitively slow for large decoders [21].

2.8 Address Transition Detector

In an asynchronous SRAM a read or a write operation is initiated by an address change or chip enable signal, whereas in synchronous SRAMs a read or write operation is initiated by the system's master clock. The terms asynchronous and synchronous relate to the memory-system interface rather than to internal chip operation.

An asynchronous SRAM features an Address-Transition Detector (ATD) that produces a pulse t_d of a controlled duration on every address transition. An ATD is typically implemented as a wide-OR gate comprising a set of one-shot circuits as shown in Figure 2.17. Every transition of address bus $A_0 - A_{N-1}$ produces a pulse on the output, which initiates a read or a write operation.

Although traditionally an asynchronous timing interface was used in standalone SRAM chips while embedded SRAMs were predominantly synchronous, recent large on-die caches seem to have reverted to an asynchronous interface with the processor core. The asynchronous interface that is employed in the dual-core

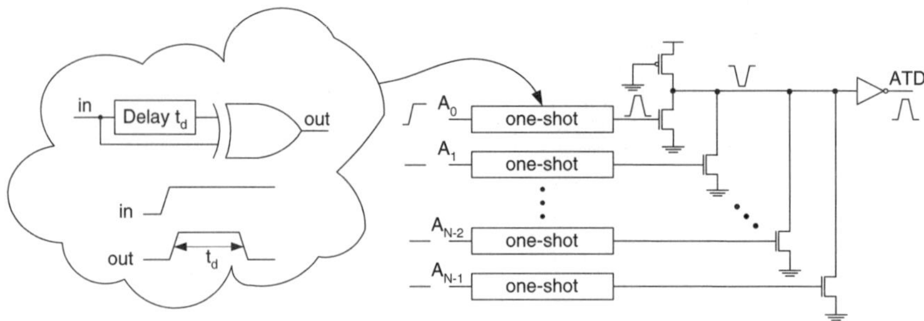

Fig. 2.17 Address Transition Detector (ATD) [21]

Itanium processor with 24 MB on-die L3 cache totalling 1.47 billion transistors [7] is reported to reduce the data latency from 8 to 5 clock cycles. The switch to an asynchronous interface in such large caches helps to combat the delay associated with the clock skew over a large die area, the latch delay, and the margins in each cycle. In addition, significant margin must be added to the SRAM cell access cycle to account for slow, marginal cells that are statistically probable in large caches. Besides the performance benefits, an asynchronous design eliminates all clock distribution and the associated latch switching power.

2.9 Timing Control Schemes

The timing control block controls precharge, word line, sense amplifier clocking and write driver activation to ensure the correct write and read operations. Technology scaling poses extra challenges for accurate timing generation. As the gate overdrive voltage is reduced with every generation of DSM process [37], V_{TH} fluctuations and process variability across the process corners are growing [38].

The key aspect of the precharged SRAM read cycle is the timing relationship between the RAM addresses, the precharge deactivation, enabling of the row and column decoders and activaton of the SA. If the asserting of the word line precedes the end of the precharge cycle, SRAM cells on the activated word line will see both the bit lines pulled high and the accessed cells may flip state. Another timing hazard may arise if the address changes before the read operation is complete i.e., when precharge is deactivated. In this case more than one SRAM cell will be discharging the bit lines which may lead to erroneous data readout. If the SA is enabled during the write operation, a "write through" can occur and the data being written will appear at the output without an intended read operation. Fundamentally, the signal path delay should match the clock path delay for correct, fast and power-efficient SRAM operation. Typically, the delay variations are dominated by the bit line delay since the minimal-size transistors in SRAM cells are more susceptible to process variations. The timing control block should provide sufficient timing margins to account for the process parameter variations with a certain number of σ that would provide a desired PPM and hence, quality level.

Basic timing control methods used in SRAMs include:

- Direct clocking (timed by the clock phase) [39]
- Delay line using a multitude of inverters to define the timing intervals [40]
- Self-timed replica (dummy) loop mimicking the signal path delay [37]
- Pipelined timing using a pipeline of registers between the SA and the data output [9]

The first method uses direct clocking of the WL and SA. It has a limited operation speed due to the larger timing margins necessary for reliable operation. The delay line allows faster operation than the direct clocking method. However, the delay of the delay loop may not track the delay variations caused by the process variations in modern nano-scaled technologies. The self-timed replica (dummy) loop method has proven more robust and precise in tracking the process variations and in maintaining tighter timing margins for faster operation. The continuing quest for speed and performance and further growth of the process variations with scaling have made the pipeline timing the method of choice in the latest SRAM designs. Below, we will discuss the delay line based, replica (dummy) loop based and the pipeline-based timing schemes in more detail.

2.9.1 Delay-Line Based Timing Control

A functional diagram of a delay-line timing loop is shown in Figure 2.18(a). A control signal *ctrl_in* sets the FSM. The timing loop is defined by the total delay through the delay elements $t_{delay1} - t_{delay\_n}$ in the FSM reset path. Typically, delay elements are based on serially-connected inverters. The delay time can be extended by using non-minimal length devices in the delay inverters or by utilizing current-starved inverters. The timing intervals formed by the delay elements $t_{delay1} - t_{delay\_n}$ with the complementary logic are used to generate the control signals for the read/write timing. The delay-line timing technique was implemented in a SRAM test chip that we will discuss in more detail in Section 5.4.2.5.

2.9.2 Replica-Loop Based Timing Control

The replica-loop based timing method provides a tighter tracking of the bit line discharge delay. A replica (dummy) column and row containing the same number of SRAM cells as in the main array are used as the reference delay elements (Figure 2.18(b)). The replica signal path mimics the capacitive loads and the associated delays of the real signal path. Therefore, it can provide more precise timing for the deactivation of the word line and the activation of the SA. Similarly to the delay-line based method, control signal *ctrl_in* sets the FSM. The output signal *out* initiates activates the word lines both in the decoded row and in the dummy row.

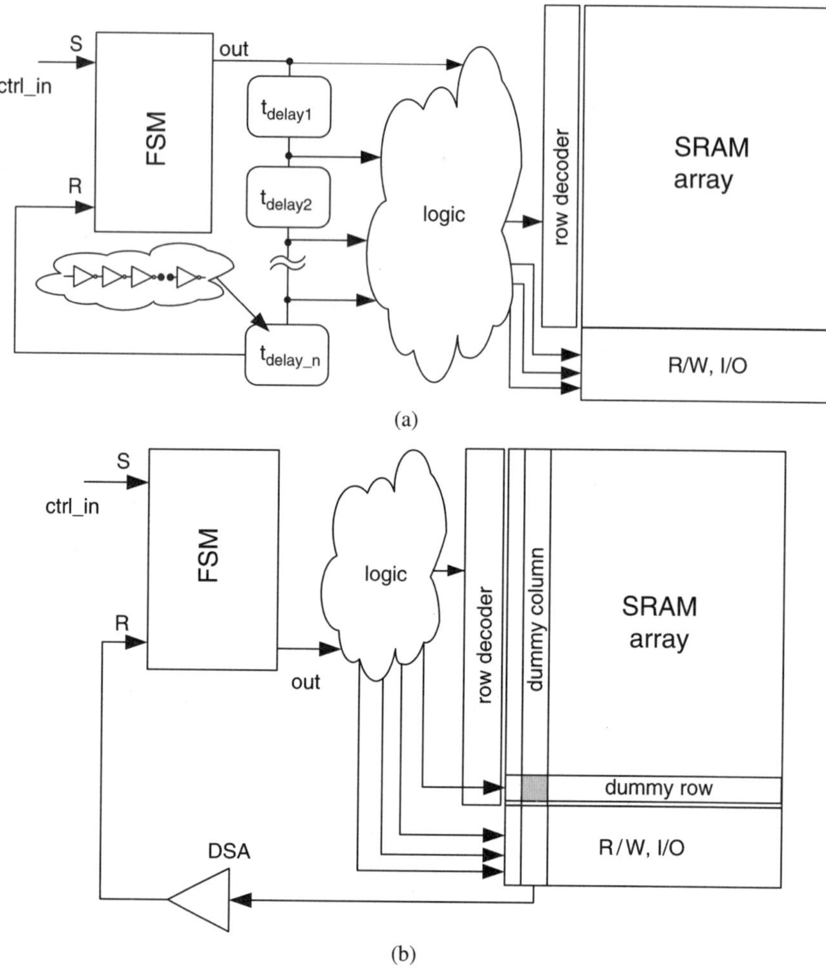

Fig. 2.18 (a) Delay chain timing loop; (b) Replica timing loop

Once the dummy bit line discharges to the dummy SA (DSA) switching threshold, the DSA flips and resets the FSM. Next, the sense amplifier enable (SAE) signal is issued and the differential voltage on the active column is amplified to the full swing. The time it takes to discharge the dummy bit line to the switching threshold of the DSA is designed to be the same as the time required for a statistically worst-case SRAM cell to develop a sufficient voltage differential on the active bit lines. Thus, bit line discharge is stopped and the SAE signal is issued as soon as a reliable read operation is guaranteed. The resulting read power savings and access time shortening make the replica-loop based timing technique popular in SRAM designs. The power dissipation overhead associated with the switching of the dummy column on each memory access is inversely proportional to the number of simultaneously accessed columns. The replica-loop based timing is explained in greater detail below.

The precision timing of the replica-based timing scheme can be achieved if the time T_{dummy} to discharge the replica bit line to the switching threshold V_{dummy} of the dummy SA equals the time T_{diff} that a regular SRAM cell needs to discharge the active bit line to achieve a sufficient differential voltage V_{diff} for reliable sensing by the bit SA. In other words,

$$T_{diff} = \frac{C_{BL\_active} V_{diff}}{I_{read}} \tag{2.6}$$

should be equal to

$$T_{dummy} = \frac{C_{BL\_dummy} V_{dummy}}{I_{dummy}}. \tag{2.7}$$

For improved sensing robustness and simplicity, an inverter is often used as a dummy SA. The bit line capacitance of the dummy column is a replica of a regular column. Combining Equations 2.6 and 2.7 and assuming that $T_{dummy} = T_{diff}$ and $C_{BL\_active} = C_{BL\_dummy} = C_{BL}$ gives:

$$\frac{C_{BL} V_{diff}}{I_{read}} = \frac{C_{BL} V_{dummy}}{I_{dummy}} \tag{2.8}$$

I_{dummy} can be expressed as:

$$I_{dummy} = \left(\frac{V_{dummy}}{V_{diff}}\right) I_{read} \tag{2.9}$$

Since the dummy SA is single-ended (inverter), its switching threshold differs from the one of the active latch-based differential SA. The switching threshold V_{dummy} of the dummy SA is chosen to be $V_{DD}/2$, whereas V_{diff} for reliable sensing is assumed to be around $V_{DD}/10$. Substituting V_{dummy} and V_{diff} in Equation 2.9 by their values gives:

$$I_{dummy} = \left(\frac{V_{DD}/2}{V_{DD}/10}\right) I_{read} = 5 I_{read} \tag{2.10}$$

In other words, in order to satisfy $T_{dummy} = T_{diff}$, the dummy bit line must be discharged with the current which is equivalent to 5X the cell read current. This condition is satisfied if the dummy bit line is discharged by simultaneous activation of five SRAM cells connected to the dummy bit line in parallel.

Once the dummy SA toggles, it resets the FSM in the control block. The reset signal disables the word line of the accessed row. The regular cells in the accessed row stop discharging the corresponding bit lines. Since $C_{BL\_active} = C_{BL\_dummy}$ and the propagation paths of the dummy loop and the regular access path are matched, V_{diff} seen by the active SA should be sufficient for reliable sensing of the stored data. The SAE signal is issued and the SA amplifies the applied V_{diff}, latches the sensed data and passes it onto the global read bus.

2.9.3 Pipelined Timing Control

Asynchronous timing schemes such as the replica-loop based timing have been popular mostly in stand-alone SRAMs. They are currently offered with access times of less than 10 ns [9]. In PC applications, synchronous SRAMs are more common. Asynchronous SRAMs are less desirable for cache applications because the rate of speeding up of their cycle time was lagging behind that of the latest high-performance processors. On the other hand, the synchronous SRAMs can address the problem of cache cycle time matching the processor cycle time. For instance, cache memories in Pentium^{TM} and PowerPC^{TM} microprocessors are built on synchronous architecture. The synchronous data buses of large SRAM arrays such as L2 and L3 cache are usually pipelined.

Pipelined SRAMs feature a pipeline of registers between the sense amplifiers and the output buffers. These registers create additional cycle latency before the first valid output can be sampled. However, the pipelined timing allows to spread the read delay (the time required to develop a sufficient bit line differential) across multiple high-speed clock cycles. The sense and the output phases of the consecutive read cycles can run in parallel at higher clock frequency compared to other timing methods. The data at the output always appears one cycle after the address selection for that data [9].

2.10 Summary

This chapter serves the background for the following chapters. In this chapter, an overview of the main SRAM building blocks was presented. We discussed four-transistor and six-transistor SRAM cells and introduced the concept of Static Noise Margin. The common building blocks of SRAM such as the sense amplifiers with the corresponding precharge and equalization circuits, several types of write drivers, row and column decoders and architecture strategies, timing control schemes including delay-based, replica-based and pipelined have been discussed.

The following chapter introduces traditional fault models and tests methods used in SRAM test.

Chapter 3
SRAM Cell Stability: Definition, Modeling and Testing

3.1 Introduction

Modern SRAMs strive to increase bit counts while maintaining low power consumption and high performance. These objectives require continuous scaling of CMOS transistors. The supply voltage mustscale down accordingly to control the power consumption and maintain the device reliability. Scaling the supply voltage and minimum transistor dimensions that are used in SRAM cells challenge the process and design engineers to achieve reliable data storage in SRAM arrays. This task is particularly difficult in large SRAM arrays that can contain millions of bits [23]. Random fluctuations in the number and location of the dopant atoms in the channel induce large threshold voltage fluctuations in scaled-down transistors [41].

Other factors affecting the repeatability of the threshold voltage and introducing V_{TH} mismatches even between the neighboring transistors in SRAM cells are the line edge roughness, the variations of the poly critical dimensions and the short channel effects [42]. SRAM stability margin or the Static Noise Margin (SNM) is projected to reduce by 4X as scaling progresses from 250 nm CMOS technology down to 50 nm technology [41, 43]. Since the stability of SRAM cells is reducing with the technology scaling, accurate estimation of SRAM data storage stability in pre-silicon design stage and verification of SRAM stability in the post-silicon testing stage are increasingly important steps in SRAM design and test flows.

This chapter will compare the static and dynamic noise margin (Section 3.2) followed by the discussion of the existing definitions of SRAM SNM (Section 3.3). Analytical expressions for calculating of the SNM of a 6T SRAM cell, the 4T cell with a resistive load and the loadless 4T SRAM cell are described in Section 3.4. The sensitivity study of the SNM to the fabrication process variations, non-catastrophic defect resistance and operating voltage is presented in Section 3.5. In Sections 3.6 and 3.7 we introduce the fault model developed to represent an SRAM cell with a reduced SNM value and the stability detection concept, respectively. The stability detection concept illustrates the principle behind the stability test techniques that we will discuss in detail in Chapter 5. We evaluated the capabilities of several march patterns to detect the stability faults in SRAM cells. The results of this study are discussed in Section 3.8.

3.2 Static Noise Margin of SRAM Cells

Noise margin can be defined using the input voltage to output voltage transfer characteristic (VTC). In general, Noise Margin (NM) is the maximum spurious signal that can be accepted by the device when used in a system while still maintaining the correct operation [44]. If the consequences of the noise applied to a circuit node are not latched, such noise will not affect the correct operation of the system and can thus be deemed tolerable. It is assumed that noise is present long enough for the circuit to react, i.e. the noise is "static" or *dc*. A Static Noise Margin is implied if the noise is a *dc* source. In case when a long noise pulse is applied, the situation is quasi-static and the noise margin asymptotically approaches the SNM (Figure 3.1) [45].

An ideal inverter tolerates a change in the input voltage (V_{in}) without any change in the output voltage (V_{out}) until the input voltage reaches the switching point. The switching point is presented in Figure 3.2(a) as $|\partial V_{out}/\partial V_{in}| = 1$. The switching point of an ideal inverter is equidistant from the logic levels. At the switching point, an ideal inverter demonstrates an absolutely abrupt change in V_{out} such that $|\partial V_{out}/\partial V_{in}| = \infty$. In other words, it has an infinite slope (gain) in the transition region. Thus, the valid logic levels (noise margins) of an ideal inverter span from the power rails to the asymptotical proximity of the transition point.

However, in a real inverter the switching point is not equidistant from the logical levels and the transition region is characterized by a finite slope $\infty > |\partial V_{out}/\partial V_{in}| > 1$, as shown in Figure 3.2(b). The finite slope $\partial V_{out}/\partial V_{in}$ in the transition region of a real inverter creates uncertainty as to what point on the VTC should be used in determining the noise margin value. This is overcome by considering a chain of inverters rather than a single inverter [44, 46]. Lohstroh et al. [46] showed that an infinite chain of inverters is similar to a flip-flop, which forms the basis of an SRAM cell.

The advantages and disadvantages of the existing noise margin definitions and their applicability to the SRAM cells noise immunity analysis are discussed in Section 3.3.

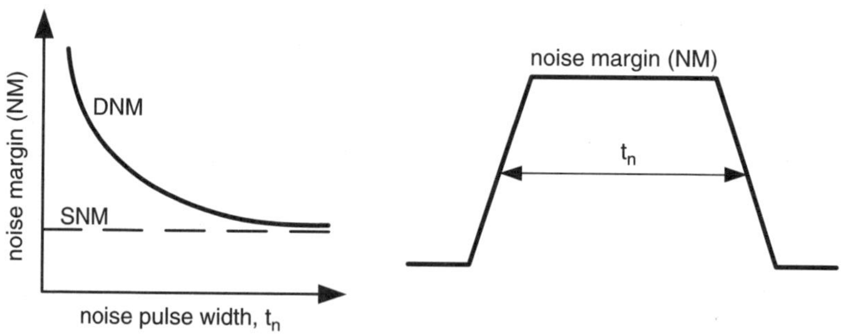

Fig. 3.1 The reduction of the Dynamic Noise Margin (DNM) with the increase of the noise pulse width t_n. For very high t_n DNM approaches its minimum (SNM)

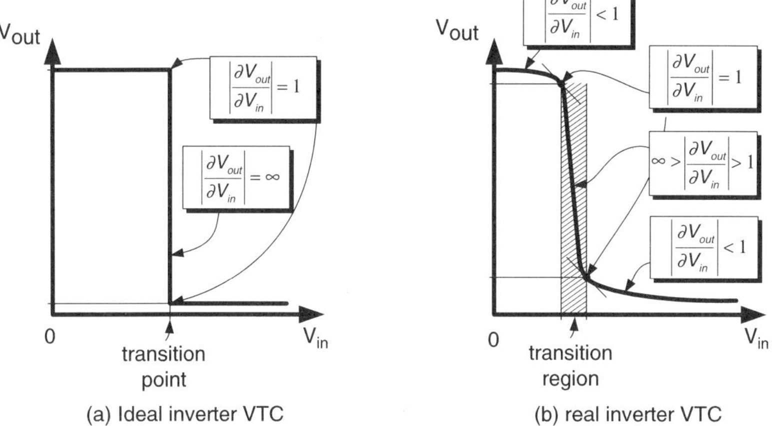

Fig. 3.2 Voltage Transfer Characteristic (VTC) of an ideal (a) and a real (b) inverter. Values of $|\partial V_{out}/\partial V_{in}|$ represent inverter gains depending on the input voltage

3.3 SNM Definitions

Several definition of the static noise margin can be found in the literature. We will describe them in detail in this section.

3.3.1 *Inverter V_{IL}, V_{IH}, V_{OL} and V_{OH}*

Several definitions of the SNM can be found in the current textbooks. One standard, commonly used SNM definition is yet to achieve universal acceptance. In textbooks [47–49], the noise margin high and noise margin low are defined as Equations 3.1 and 3.2, respectively:

$$NM_H = V_{OH} - V_{IH} \tag{3.1}$$

$$NM_L = V_{IL} - V_{OL} \tag{3.2}$$

where V_{IL} is the maximum input voltage level recognized as logical "0", V_{IH} is the minimum input voltage level recognized as a logical "1", V_{OL} is the maximum logical "0" output voltage, V_{OH} is the minimum logical "1" output voltage as illustrated in Figure 3.3.

Any inverter transfer curve, which falls into the shaded area, will have noise margins at least as good as given by the equations above. This approach specifies the compatibility of the logic levels of circuits in the same logic family. To provide the correct signal interaction without the need to employ the level conversion circuitry, the output logic levels of one circuit must be compatible with the input logic levels of the next stage. For instance, input voltage in the range of $V_{IL} < V_{in} < V_{IH}$ may not be properly recognized by the gate and may cause a logic error.

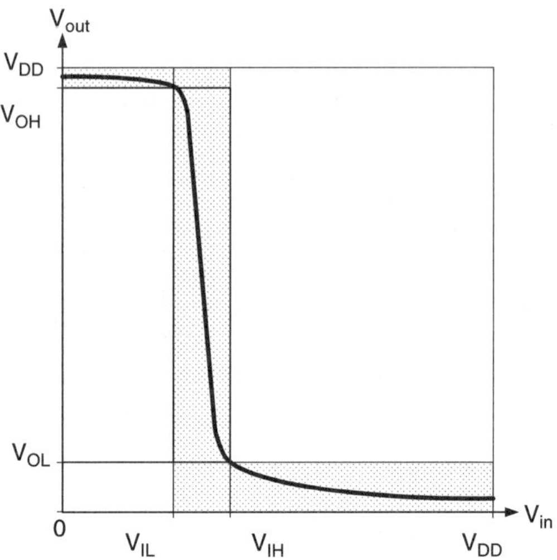

Fig. 3.3 Definition of V_{IL}, V_{IH}, V_{OL} and V_{OH} in Equations 3.1 and 3.2. The inverter transfer curve must lie within the shaded region to be a member of the logic family

3.3.2 Noise Margins NM_H and NM_L with V_{OL} and V_{OH} Defined as Stable Logic Points

Figure 3.4 represents one of the ways to define V_{OH} and V_{OL}. In this approach, V_{OH} and V_{OL} are represented as the stable voltage states of a bistable inverter pair. The resulting values of NM_L and NM_H are larger than for other definitions. Defining V_{OH} and V_{OL} like this runs into trouble with the basic and simple concepts discussed in connection with logic level definitions depicted in Figure 3.3. The transfer characteristic does not lie within the required shaded area and thus cannot represent a set of valid logic level definitions from which any meaningful noise margins can be calculated. Thus, this approach must be rejected as a valid noise margin approach [44, 50].

3.3.3 Noise Margins NM_H and NM_L with V_{OL} and V_{OH} Defined as −1 Slope Points

Figure 3.5 represents another way to define the noise margins. In this approach V_{OH} and V_{OL} are represented as the stable points where the $dVout/dVin = -1$ of a bistable inverter pair and coincide with V_{IL} and V_{IH} respectively. That makes more sense since V_{OL} is defined as the maximum output voltage level with the gate at "0" logical state, and V_{OH} is the minimum output voltage with "1" logical level. The resulting values of NM_L and NM_H are smaller than those produced by using the noise margin definition in Section 3.3.2.

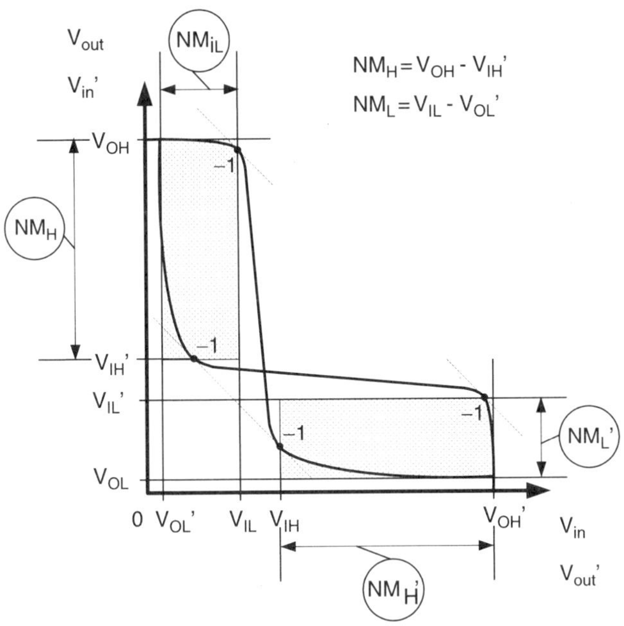

Fig. 3.4 Graphical representation of SNM with V_{OH} and V_{OL} in Equations 3.1 and 3.2 as stable logic state points of a bistable inverter pair

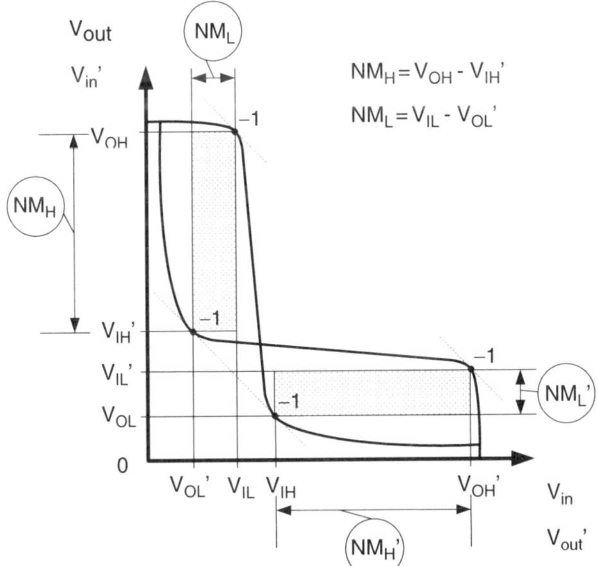

Fig. 3.5 Graphical representation of SNM with V_{OH} and V_{OL} in Equations 3.1 and 3.2 as -1 slope points of a bistable inverter pair

This technique represents more of a legacy from considering the noise margins of a inverter chain than a bistable inverter pair, when the mirrored transfer characteristics were not considered on the same coordinate system. The V_{OH} and V_{OL} levels have theoretical justification in this approach. The sum of NM_L and NM_H is maximum if the V_{OH} and V_{OL} points are chosen to be at the -1 slope points [46]. However, the application of this approach imposes some restrictions on the shape of the transfer curves as it maximizes the sum of the noise margins and not the individual noise margins. This theoretically can lead to a situation, when the maximum sum occurs at points where one of the noise margins is zero or even negative [50]. For the transfer curves of an SRAM cell, however, this criterion produces reliable results. Another advantage of this approach is that it allows analytical calculation of the noise margins of an SRAM cell.

3.3.4 SNM as a Side of the Maximum Square Drawn Between the Inverter Characteristics

The approach was first described by Hill [44] in 1968. An important advantage of this method is that it can be automated using a DC circuit simulator, which to a great degree extends its practical usefulness. In this approach an SRAM cell is presented as two equivalent inverters with the noise sources inserted between the corresponding inputs and outputs (Figure 3.6). Both series voltage noise sources (V_n) have the same value and act together to upset the state of the cell, i.e. they have an "adverse" polarity to the current state of each inverter of the cell. Applying the adverse noise sources polarity represents the worst-case equal noise margins [50]. This method is only applicable to circuits with $R_{in} \gg R_{out}$, and CMOS inverters of an SRAM cell comply with this condition.

Having two adverse noise sources applied to the input of each inverter of an SRAM cell makes the value of the obtained SNM to be the worst-case SNM as shown in Figure 3.7 [46]. In contrast, the best-case SNM would be obtained if only one noise source was applied or the polarities of the noise sources were not adverse. However, one is rarely interested in such an idealized case.

Figure 3.8 shows the superimposed normal inverter transfer curve of a read-accessed 6T SRAM cell and its mirrored with respect to $x = y$ line counterpart in a

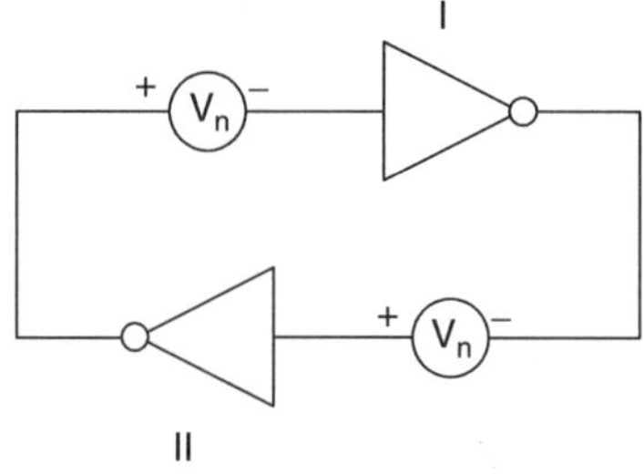

Fig. 3.6 Flip-flop with two noise sources with adverse polarities

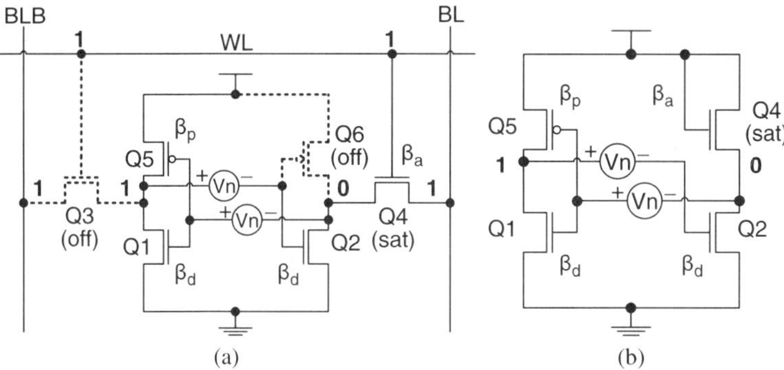

Fig. 3.7 Read-accessed SRAM cell with inserted adverse polarity static noise sources V_n (a) and its equivalent circuit (b)

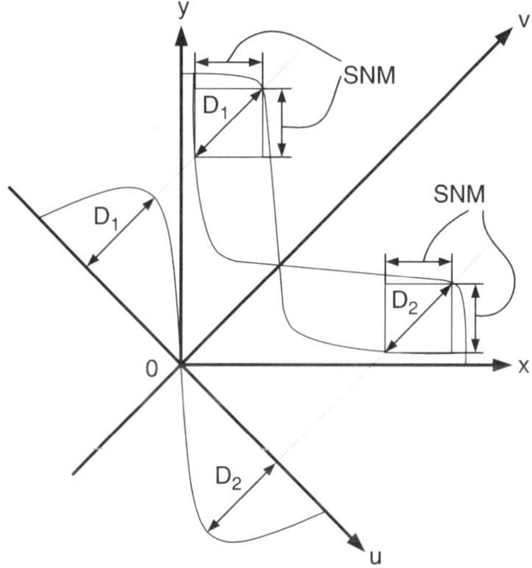

Fig. 3.8 SNM estimation based on "maximum squares" in a 45° rotated coordinate system. The voltage transfer characteristics (VTCs) of both inverters comprising an SRAM cell are ideally symmetrical

$x-y$ coordinate system. The $u-v$ system of coordinates is rotated counter clockwise by 45° around the same origin with respect to $x-y$ system. This is a convenient arrangement. Since by knowing the diagonals of the maximum embedded squares we can calculate the sides. The v axis is parallel to the sought diagonals. The dashed curve in the $u-v$ coordinate system represents the subtraction of the normal and mirrored inverter transfer curves in the $x-y$ coordinate system. Since the squares have maximum size when the lengths of their diagonals D_1 and D_2 are maximum, the extremes of this curve correspond to the diagonals of the maximum embedded squares. Generally, due to the process spread, $D_1 \neq D_2$. Suppose that $D_1 > D2$. Then $D_1/\sqrt{2}$ yields the SNM of the flip-flop.

The above algorithm can be expressed mathematically in the following manner. Assume that the normal and mirrored inverter characteristics are defined as functions $y = F_1(x)$ and $y = F_2'(x)$, where the latter is the mirrored version of $y = F_2(x)$. To find F_1 in terms of u and v, the $x - y$ coordinate system has to be transformed as follows [51]:

$$x = \frac{1}{\sqrt{2}}u + \frac{1}{\sqrt{2}}v \tag{3.3}$$

$$y = -\frac{1}{\sqrt{2}}u + \frac{1}{\sqrt{2}}v \tag{3.4}$$

substitution of Equations 3.3 and 3.4 in $y = F_1(x)$ gives:

$$v = u + \sqrt{2}F_1\left(\frac{1}{\sqrt{2}}u + \frac{1}{\sqrt{2}}v\right) \tag{3.5}$$

For F_2', first F_2 is mirrored in the $x - y$ system with respect to line $x = y$ (v axis) and then transformed using the same technique as in Equations 3.3 and 3.4 but with x and y interchanged, which produce:

$$x = -\frac{1}{\sqrt{2}}u + \frac{1}{\sqrt{2}}v \tag{3.6}$$

$$y = \frac{1}{\sqrt{2}}u + \frac{1}{\sqrt{2}}v \tag{3.7}$$

substitution of Equations 3.6 and 3.7 in $y = F_2(x)$ gives:

$$v = -u + \sqrt{2}F_2\left(-\frac{1}{\sqrt{2}}u + \frac{1}{\sqrt{2}}v\right) \tag{3.8}$$

Equations 3.5 and 3.8 explicitly express v as a function of u. Solutions of Equations 3.5 and 3.8 can be found using a standard HSPICE-like DC circuit simulator by translating the equations into circuits with voltage-dependent voltage sources in a feedback loop [51].

The solutions of Equations 3.5 and 3.8 are represented by v_1 and v_2 in Figure 3.9. The difference between the two solutions v_1 and v_2 is represented by the sine-like curve in Figure 3.8. The absolute values of the extremes of this curve (D) where $dD/du = 0$ represent the lengths of the diagonals of the squares embedded between the direct and mirrored SRAM flip-flop inverter curves. Multiplication of the smaller of the two by $1/\sqrt{2}$ yields the worst-case SNM of an SRAM cell.

3.4 Analytical Expressions for SNM Calculation

Analytical expressions for SNM calculations can help to estimate the impact of various process parameters on the SRAM cell stability. Another possible application of analytical expressions is optimizing the SRAM cell design to the requirements of a specific application such as high performance, low power or low voltage.

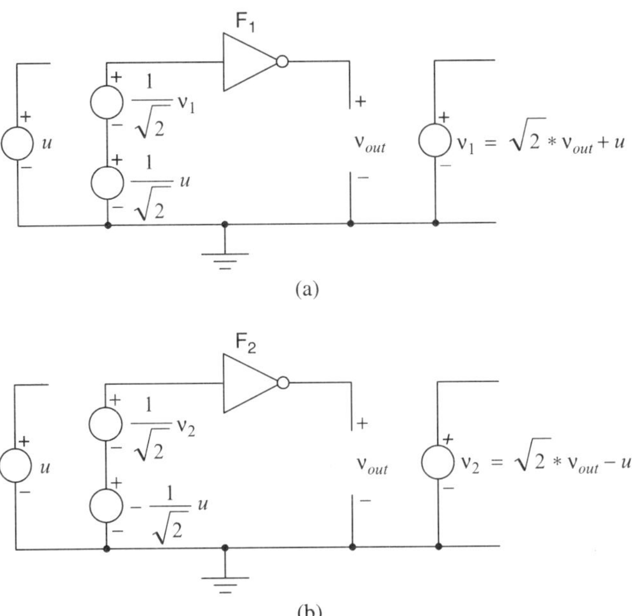

Fig. 3.9 Circuit implementation of Equations 3.5(a) and 3.8(b) for finding the diagonal of the square embedded between the direct and mirrored SRAM flip-flop inverter curves

We will consider the SNM analytical models of three types of SRAM cells: a four-transistor (4T) cell with a resistor load; a six-transistor 6T full CMOS cell; and a four-transistor (4) loadless cell with PMOS access transistors. Analytical expressions for a 4T SRAM cell with a polysilicon resistor load and a 6T CMOS SRAM cell (Sections 3.4.1 and 3.4.2 respectively) were obtained using the basic MOS model equation with constant V_{TH} and neglecting second-order effects such as mobility reduction and velocity saturation [51]. The analytical expression for a loadless 4T SRAM cell presented in Section 3.4.4.2 is derived using the alpha-power law for MOS transistor current equations.

3.4.1 Analytical SNM Expression for a 4T SRAM Cell with Polysilicon Resistor Load

The long-channel MOS equations are:

$$I_{D(sat)} = \frac{1}{2}\beta(V_{GS} - V_{TH})^2 \tag{3.9}$$

and

$$I_{D(lin)} = \beta\left(V_{GS} - V_{TH} - \frac{1}{2}V_{DS}\right) \tag{3.10}$$

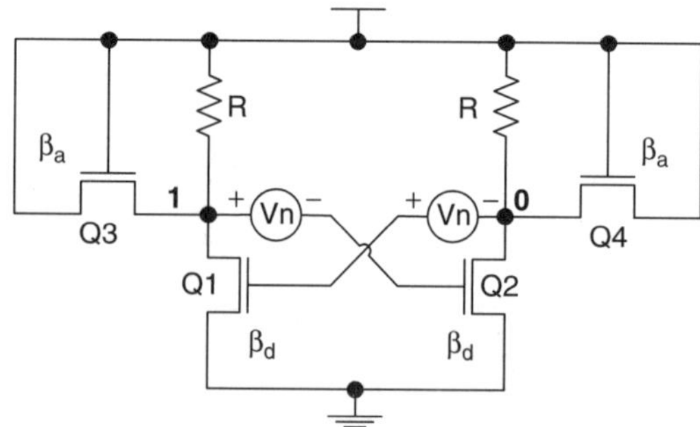

Fig. 3.10 A four-transistor SRAM cell with R-load in a read-accessed mode

For analysis of the circuit in Figure 3.10, we assume that $Q1$, $Q3$ and $Q4$ are saturated, whereas $Q2$ is in the linear mode.

The voltage gain of each inverter is given as $\sqrt{r}$, where $r = \beta_{driver}/\beta_{access}$. The loop gain r must be larger than unity for the sufficient flip-flop operation [46]. Equating the drain currents in the cell inverters we find:

$$I_{DQ3} = I_{DQ1} = V_{GS3} - V_{TH} = \sqrt{r}(V_{GS1} - V_{TH}) \tag{3.11}$$

$$I_{DQ4} = I_{DQ2} = (V_{GS4} - V_{TH})^2 = 2r\left(V_{GS2} - V_{TH} - \frac{1}{2}V_{DS2}\right) \tag{3.12}$$

The Kirchhoff equations for $Q1$, $Q3$ and $Q4$ of the 4T SRAM cell with a resistive load are:

$$V_{GS1} = V_n + V_{DS2} \tag{3.13}$$

$$V_{GS3} = V_{DD} - V_{GS2} - V_n \tag{3.14}$$

$$V_{GS4} = V_{DD} - V_{DS2} \tag{3.15}$$

Substituting Equations 3.13–3.15 into Equations 3.11 and 3.12 yields:

$$V_{DD} - V_{GS2} - V_n - V_{TH} = \sqrt{r}(V_n + V_{DS2} - V_{TH}) \tag{3.16}$$

$$(V_{DD} - V_{DS2} - V_{TH})^2 = 2rV_{DS2}\left(V_{GS2} - V_{TH} - \frac{1}{2}V_{DS2}\right) \tag{3.17}$$

Eliminating V_{GS2} from Equations 3.16 and 3.17 and simplifying results in a quadratic equation

$$aV_{DS2}^2 - bV_{DS2} + c = 0 \tag{3.18}$$

where

$$\begin{aligned} a &= 1 + r + 2r^{3/2} \\ b &= -2\left(V_s(r+1) + r(\sqrt{r}-1)V_{TH} - r(\sqrt{r}+1)V_n\right) \\ c &= V_s^2 \\ V_s &= V_{DD} - V_{TH} \end{aligned} \tag{3.19}$$

Now we apply the condition of coinciding roots [46] to find the SNM using Equations 3.18 and 3.19. For Equations 3.18 this means a double root, which requires

$$b^2 = 4ac$$

or

$$b = -2\sqrt{ac} \tag{3.20}$$

since $b < 0$. Substituting Equations 3.19 and solving for V_n yields the SNM of a 4T cell with a resistive load [51]:

$$SNM_{4T_R} = \frac{\sqrt{r}-1}{\sqrt{r}+1}V_{TH} + \frac{r+1-\sqrt{2r^{3/2}+r+1}}{r(\sqrt{r}+1)}V_s. \tag{3.21}$$

3.4.2 Analytical SNM Expression for a 6T SRAM Cell

Employing the same long-channel MOS current equations as in Section 3.4.1 for the circuit in Figure 3.11 and assuming that transistors $Q1$ and $Q4$ are saturated and transistors $Q2$ and $Q5$ are in the linear mode:

$$(V_{GS1} - V_{TH})^2 = \frac{2q}{r}V_{DS5}\left(V_{GS5} - V_{TH} - \frac{1}{2}V_{DS5}\right) \tag{3.22}$$

$$(V_{GS4} - V_{TH})^2 = 2rV_{DS2}\left(V_{GS2} - V_{TH} - \frac{1}{2}V_{DS2}\right) \tag{3.23}$$

where V_{TH} of NMOS transistors in the cell is assumed V_{TH} of PMOS transistors and $q = \beta_p/\beta_a$, $r = \beta_d/\beta_a$.

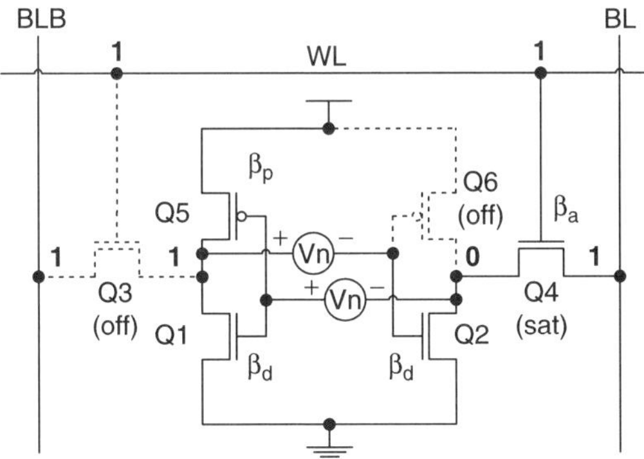

Fig. 3.11 A six-transistor full CMOS SRAM cell in a read-accessed mode

The Kirchhoff equations for the 6T SRAM cell are:

$$V_{GS1} = V_n + V_{DS2} \tag{3.24}$$

$$V_{DS5} = V_{DD} - V_n - V_{GS2} \tag{3.25}$$

$$V_{GS5} = V_{DD} - V_n - V_{DS2} \tag{3.26}$$

and

$$V_{GS4} = V_{DD} - V_{DS2} \tag{3.27}$$

Substituting Equations 3.24–3.27 into Equations 3.22 and 3.23 yields:

$$\begin{aligned}(V_{DS2} + V_n - V_T)^2 &= \tfrac{q}{r}\,(V_{DD} - V_n - V_{GS2}) \times \\ &\times (V_s - V_{TH} - V_n - 2V_{DS} + V_{GS2})\end{aligned} \tag{3.28}$$

$$(V_s - V_{DS2})^2 = 2rV_{DS2}\left(V_{GS2} - V_{TH} - \frac{1}{2}V_{DS2}\right) \tag{3.29}$$

where $V_s = V_{DD} - V_{TH}$

Eliminating V_{GS2} and V_{DS2} from Equations 3.28 and 3.29 yields a fourth-degree equation. Assuming local linearity of the transfer characteristic on inverter *Q2–Q4* around its operating point where *Q2* is in the linear region, it can be simplified as [51]:

$$V_{DS2} = V_0 - kV_{GS2} \tag{3.30}$$

where

$$V_r = V_s - \left(\frac{r}{r+1}\right)V_{TH} \tag{3.31}$$

$$k = \left(\frac{r}{r+1}\right)\left(\sqrt{\frac{r+1}{r+1-V_s^2/V_r^2}} - 1\right) \tag{3.32}$$

$$V_0 = kV_s + \left(\frac{1+r}{1+r+r/k}\right)V_r \tag{3.33}$$

After eliminating V_{DS2} from Equations 3.28 and 3.30 and simplification, we obtain:

$$X^2\left(1 + 2k + \frac{r}{q}k^2\right) + 2X\left(\frac{r}{q}kA + A + V_{TH} - V_s\right) + \frac{r}{q}A^2 = 0 \tag{3.34}$$

where

$$\begin{aligned}X &= V_{DD} - V_n - V_{GS2}\\ A &= V_0 + (k+1)V_n - kV_{DD} - V_{TH}\end{aligned} \tag{3.35}$$

Similarly to the derivation presented in Section 3.4.1, the double-root stability criterion was applied to Equation 3.34. Next, when Equation 3.35 is substituted, Equation 3.34 can be solved for the dc disturbance (transistor offsets and mismatches and variations in operating conditions) represented by V_n to obtain the SNM:

$$SNM_{6T} = V_{TH} - \left(\frac{1}{k+1}\right) \times \\ \times \left(\frac{V_{DD} - \frac{2r+1}{r+1} V_{TH}}{1 + \frac{r}{k(r+1)}} - \frac{V_{DD} - 2V_{TH}}{1 + k\frac{r}{q} + \sqrt{\frac{r}{q}\left(1 + 2k + \frac{r}{q}k^2\right)}} \right) \tag{3.36}$$

3.4.3 Conclusions from SNM Analytical Expressions

The analysis of Equations 3.21 and 3.36 shows that the SNM depends only on V_{TH}, V_{DD} and β cell ratios and not on the absolute values of β. Both the SNM_{4T_R} and SNM_{6T} increase with r. $SNM_{6T} > 0$ for all $r > 0$, whereas $SNM_{4T_R} = 0$ for $r = 1$.

To maximize the SNM, the cell transistor ratios β must be selected such that $r = \beta_d/\beta_a = max$ and also (in the case of a 6T CMOS SRAM cell) $q/r = \beta_p/\beta_d = max$. On the other hand, the cell ratios are constrained by the minimum area and reliable write operation requirements.

For particular values of q and r, the SNM_{6T} is independent of V_{DD} variations. This is due to the coefficients of V_{DD} in Equation 3.36 having opposite signs. Modifying q or r will then result in either a positive or a negative dependence of SNM_{6T} on V_{DD}. This property means that a particular cell stability behavior as a function of V_{DD} can be obtained by selecting specific values of q and r. However, for the 4T cell with a resistive load, the SNM_{4T_R} will always decrease with decreasing V_{DD}.

Larger V_{TH} of the cell transistors will produce larger SNM_{4T_R} and SNM_{6T}. Since V_{TH} decreases with temperature, the SNM will also decrease with temperature.

Analytical expressions presented in Sections 3.4.1 and 3.4.2 show satisfactory matching with the simulation results. The deviations of the analytical results occur due to the fact that these expressions did not account for the short-channel effects. As a general observation, the 6T SRAM cell requires a significantly smaller cell ratio than the 4T cell with a resistive load to achieve the same SNM for smaller V_{DD}. Thus, the advantage of the 4T cells over the 6T cells begins to disappear in sub-micron technologies featuring reduced supply voltages that are needed to maintain constant electric field scaling.

3.4.4 Analytical SNM Expression for a Loadless 4T SRAM Cell

In this section we derive an analytical SNM expression for a loadless 4T SRAM cell using the alpha-power law model [52].

3.4.4.1 Alpha-Power Law Model

Shockley's MOSFET model, represented by Equation 3.37 was used to analytically calculate circuit parameters for long-channel transistors. However, the Shockley model is increasingly inaccurate in describing the behavior of the modern short-channel transistors. Short-channel effects, such as the carrier velocity saturation, must be taken into account for accurate analytical characterization of sub-micron MOS FETs.

$$I_D = \begin{cases} 0, \text{ for } V_{GS} \leq V_{TH} & - \text{ cutoff} \\ K((V_{GS} - V_{TH})V_{DS} - 0.5V_{DS}^2), \text{ for } V_{DS} \leq V_{DSAT} & - \text{ linear} \\ 0.5K(V_{GS} - V_{TH})^2, \text{ for } V_{DS} \geq V_{DSAT} & - \text{ saturation} \end{cases} \tag{3.37}$$

where $V_{DSAT} = V_{GS} - V_{TH}$ is the drain saturation voltage, V_{TH} is the threshold voltage, K is the drivability factor (Equation 3.38) defined as follows:

$$K = \mu(\varepsilon_{ox}/t_{ox})(W/L_{EFF}) \tag{3.38}$$

where μ is the effective mobility, ε_{ox} and t_{ox} are the gate oxide dielectric constant and thickness respectively and W and L_{EFF} are the transistor channel width and effective channel length, respectively.

In scaled-down transistors, Shockley's square-law dependence does not hold. The shift of V_{DSAT} and discrepancies in the saturation region called for the Alpha-Power Law (APL) proposed in [53]. Drain current in the APL is then proportional to $(V_{GS} - V_{TH})^{\alpha}$, where α is the velocity saturation index. While in the Shockley model $\alpha = 2$, the measured α values can range from one to two. The APL model is defined as shown in Equation 3.39.

$$I_D = \begin{cases} 0, \text{ for } V_{GS} \leq V_{TH} & - \text{ cutoff} \\ (I'_{D0}/V'_{D0})V_{DS}, \text{ for } V_{DS} \leq V'_{D0} & - \text{ linear} \\ I'_{D0}, \text{ for } V_{DS} \geq V'_{D0} & - \text{ saturation} \end{cases} \tag{3.39}$$

where

$$I'_{D0} = I_{D0}\left(\frac{V_{GS} - V_{TH}}{V_{DD} - V_{TH}}\right)^{\alpha} \tag{3.40}$$

$$V'_{D0} = V_{D0}\left(\frac{V_{GS} - V_{TH}}{V_{DD} - V_{TH}}\right)^{\alpha/2} \tag{3.41}$$

V_{D0} is the drain saturation voltage at $V_{GS} = V_{DD}$ and I_{D0} is the drain current at $V_{GS} = V_{DS} = V_{DD}$.

The MOSFET drain current equations can also be rewritten as in [54]:

$$I_D = \begin{cases} 0, \text{ for } V_{GS} \leq V_{TH} & - \text{ cutoff region} \\ K_L(V_{GS} - V_{TH})^{\alpha/2}V_{DS}, \text{ for } V_{DS} < V'_{D0} & - \text{ linear region} \\ K_S(V_{GS} - V_{TH})^{\alpha}, \text{ for } V_{DS} \geq V'_{D0} & - \text{ saturation region} \end{cases} \tag{3.42}$$

where

$$K_L = \frac{I_{D0}}{V_{D0}(V_{DD} - V_{TH})^{\alpha/2}} \tag{3.43}$$

$$K_S = \frac{I_{D0}}{(V_{DD} - V_{TH})^{\alpha}} \tag{3.44}$$

We will use Equation 3.42 to derive an analytical expression for the SNM calculation in four-transistor loadless SRAM cells (Section 3.4.4.2).

3.4.4.2 Analytical SNM Expression Derivation

Assuming that both inverters comprising a four-transistor loadless SRAM cell are equivalent, we used the following equivalent circuits to derive the SNM expression (Figure 3.12). Since we are interested in the worst-case SNM, we will consider the cell in the read-accessed mode, i.e. with the activated word line. In the case of a four-transistor loadless SRAM cell, which is using PMOS transistors as both the access and the load, the read-accessed mode corresponds to $V_{WL} = 0$, i.e. when the gate of $Q4$ is grounded. Since many of the parameters in the APL model are technology-dependent, we will differentiate between n and p transistors as well as between the linear and saturated modes of the transistors. For instance, the velocity saturation index α, the threshold voltage V_{TH} and the saturation voltage V_{D0} will vary with the transistor type and operating mode. We will explain this in more detail in Section 3.4.4.6.

3.4.4.3 Finding V_{OH} and V_{IL}

Throughout the derivation we will use the following convention for $Q2$ and $Q4$ in Figure 3.12(b) and (c). $Q2$ will be represented by a subscript n and $Q4$ – by a subscript p.

Analyzing Figure 3.12(a) we can conclude that the NMOS transistor $Q2$ in Figure 3.12(b) is in the saturation region (denoted as *sat*) and PMOS transistor $Q4$ is in the linear region (*lin*). Using Equation 3.42 and the KCL, we can equate the I_D equations for NMOS transistor in the saturation region and for the PMOS transistor in the linear mode. Using the APL model results in:

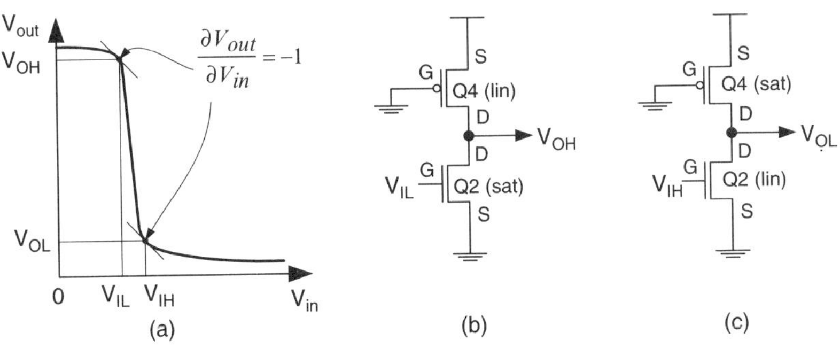

Fig. 3.12 Definitions of V_{OH}, V_{IL}, V_{IH} and V_{OL} (a); Equivalent circuit of a 4T loadless SRAM half-cell showing the transistor modes for V_{IL} and V_{OH} (b) and for V_{IH} and V_{OL} (c)

$$K_{s\_n(sat)}\left(V_{GS\_n}-V_{TH\_n(sat)}\right)^{\alpha\_n(sat)}= \\ =-K_{l\_p(lin)}\left(\left|V_{GS\_p}-V_{TH\_p(lin)}\right|\right)^{\alpha\_p(lin)/2}V_{DS\_p} \quad (3.45)$$

Note that $V_{GS\_P}=-V_{DD}$, $V_{GS\_N}=V_{IL}$, $V_{DS\_N}=V_{OH}$ and $V_{DS\_P}=V_{OH}-V_{DD}$. Due to $V_{D0}<0$ for PMOS transistor $Q4$, $K_p<0$. Then, Equation 3.45 can then be rewritten as:

$$K_{s\_n(sat)}\left(V_{IL}-V_{TH\_n}\right)^{\alpha\_n(sat)}= \\ =-K_{l\_p(lin)}\left(\left|-V_{DD}-V_{TH\_p(lin)}\right|\right)^{\alpha\_p(lin)/2}\left(V_{OH}-V_{DD}\right) \quad (3.46)$$

V_{OH} can be expressed as:

$$V_{OH}=-\frac{K_{s\_n(sat)}\left(V_{IL}-V_{TH\_n(sat)}\right)^{\alpha\_n(sat)}}{K_{l\_p(lin)\left(\left|-V_{DD}-V_{TH\_p(lin)}\right|\right)^{\alpha\_p(lin)/2}}}+V_{DD} \quad (3.47)$$

We defined V_{OH} as the point where $\delta V_{OH}/\delta V_{IL}=-1$:

$$\frac{\partial V_{OH}}{\partial V_{IL}}=\begin{bmatrix}-\frac{K_{s\_n(sat)}}{K_{l\_p(lin)}}\left(V_{IL}-V_{TH\_p(lin)}\right)^{\alpha\_n(sat)}K_{l\_p(lin)}\times \\ \times\left(\left|-V_{DD}-V_{TH\_p(lin)}\right|\right)^{-\alpha\_p(lin)/2}\end{bmatrix}'=-1 \quad (3.48)$$

Solving Equation 3.48 for V_{IL}, we obtain:

$$V_{IL}=\left(\frac{K_{l\_p(lin)}}{K_{s\_n(sat)}}\frac{1}{\alpha_{n(sat)}}\right)^{\frac{1}{\alpha_{n(sat)}-1}}\times \\ \times\left(\left|-V_{DD}-V_{TH\_p(lin)}\right|\right)^{\frac{\alpha_{p(lin)}}{2\left(\alpha_{n(sat)}-1\right)}}+V_{TH\_n(sat)} \quad (3.49)$$

3.4.4.4 Finding V_{OL} and V_{IH}

Now, let us consider the point where $V_{in}=V_{IH}$ and $V_{out}=V_{OL}$ on the VTC of the four-transistor loadless SRAM half-cell. Analyzing Figure 3.12 (a) we can conclude that the NMOS transistor $Q2$ in Figure 3.12 (c) is in the linear region and the PMOS transistor $Q4$ is in the saturation region. Using Equation 3.42 and the KCL, we can equate the I_D equations for the NMOS transistor in the linear mode and for the PMOS transistor in the saturation mode. Using the APLM results in:

$$K_{l\_n(lin)}\left(V_{GS\_n}-V_{TH\_n(lin)}\right)^{\alpha\_n(lin)/2}V_{DS\_n}= \\ =K_{s\_p(sat)}\left(V_{GS\_p}-V_{TH\_p(sat)}\right)^{\alpha\_p(sat)} \quad (3.50)$$

Note that $V_{GS\_P}=-V_{DD}$, $V_{GS\_N}=V_{IH}$, $V_{DS\_N}=V_{OL}$. Equation 3.50 can be rewritten as:

$$K_{l\_n(lin)}\left(V_{IH}-V_{TH\_n(lin)}\right)^{\alpha\_n(lin)/2}V_{OL}= \\ =K_{s\_p(sat)}\left(\left|-V_{DD}-V_{TH\_p(sat)}\right|\right)^{\alpha\_p(sat)} \quad (3.51)$$

V_{OL} can be expressed as:

$$V_{OL} = \frac{K_{s\_p(sat)}\left(\left|-V_{DD} - V_{TH\_p(sat)}\right|\right)^{\alpha\_p(sat)}}{K_{l\_n(lin)}\left(V_{IH} - V_{TH\_n(lin)}\right)^{\alpha\_n(lin)/2}} \tag{3.52}$$

We defined V_{OL} as the point where $\delta V_{OL}/\delta V_{IH} = -1$:

$$\frac{\partial V_{OL}}{\partial V_{IH}} = \begin{bmatrix} \frac{K_{s\_p(sat)}}{K_{l\_n(lin)}}\left(\left|-V_{DD} - V_{TH\_p(sat)}\right|\right)^{\alpha\_p(sat)} \times \\ \times\left(V_{IH} - V_{TH\_n(lin)}\right)^{-\alpha_n(lin)/2} \end{bmatrix}' = -1 \tag{3.53}$$

Solving Equation 3.48 for V_{IH}, it can be expressed as:

$$V_{IH} = \left(\frac{K_{l\_n(lin)}}{K_{s\_p(sat)}}\frac{2}{\alpha_{n(sat)}}\right)^{-\frac{2}{\alpha_{n(sat)}+2}} \times \\ \times\left(\left|-V_{DD} - V_{TH\_p(sat)}\right|\right)^{\frac{2\alpha_{p(sat)}}{\alpha_{n(lin)}+2}} + V_{TH\_n(lin)} \tag{3.54}$$

3.4.4.5 SNM Expression for 4T Loadless SRAM Cell

Following Equations 3.1 and 3.2 and substituting V_{OH}, V_{IL}, V_{OL} and V_{IH} from Equations 3.47, 3.47, 3.52 and 3.54 respectively, we can find the Noise Margin High (NM_H):

$$NM_H = V_{OH} - V_{IH} = V_{DD} - \frac{K_{s\_n(sat)}\left(V_{IL} - V_{TH\_n(sat)}\right)^{\alpha\_n(sat)}}{K_{l\_p(lin)}\left(\left|-V_{DD} - V_{TH\_p(lin)}\right|\right)^{\alpha\_p(lin)/2}} - \\ -\left(\frac{K_{l\_n(lin)}}{K_{s\_p(sat)}}\frac{2}{\alpha_{n(sat)}}\right)^{-\frac{2}{\alpha_{n(sat)}+2}}\left(\left|-V_{DD} - V_{TH\_p(sat)}\right|\right)^{\frac{2\alpha_{p(sat)}}{\alpha_{n(lin)}+2}} + V_{TH\_n(lin)}, \tag{3.55}$$

and the Noise Margin Low (NM_L):

$$NM_L = V_{IL} - V_{OL} = \\ = \left(\frac{K_{l\_p(lin)}}{K_{s\_n(sat)}}\frac{1}{\alpha_{n(sat)}}\right)^{\frac{1}{\alpha_{n(sat)}-1}}\left(\left|-V_{DD} - V_{TH\_p(lin)}\right|\right)^{\frac{\alpha_{p(lin)}}{2\left(\alpha_{n(sat)}-1\right)}} + \\ +V_{TH\_n(sat)} - \frac{K_{s\_p(sat)}\left(\left|-V_{DD} - V_{TH\_p(sat)}\right|\right)^{\alpha\_p(sat)}}{K_{l\_n(lin)}\left(V_{IH} - V_{TH\_n(lin)}\right)^{\alpha\_n(lin)/2}}. \tag{3.56}$$

NM_H and NM_L represent the sides of a rectangle embedded between the two VTCs of the half-cells (Figure 3.13). For analytical calculations, we can express the SNM of a loadless four-transistor SRAM cell as the diagonal of the rectangle with the sides equal to NM_H and NM_L. The final expression can be presented as:

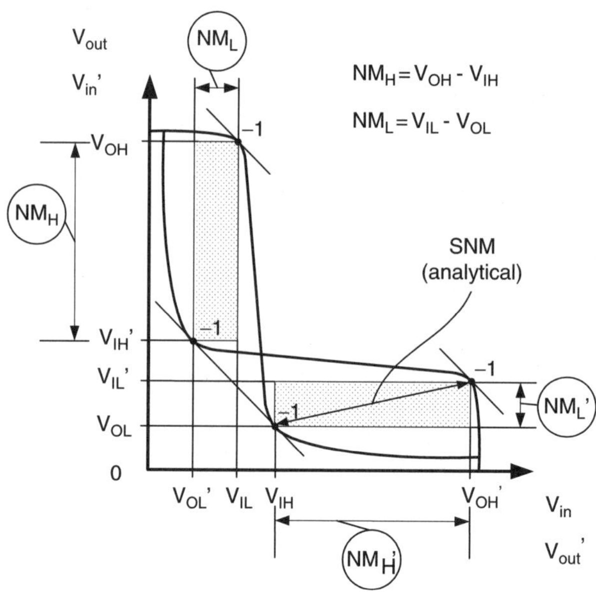

Fig. 3.13 SNM definition used in the analytical SNM expression for a four-transistor loadless SRAM cell (Equation 3.57)

$$SNM_{4T\_loadless\_SRAM} = \sqrt{NM_H^2 + NM_{LH}^2} =$$

$$= \sqrt{\begin{array}{l} \left(\begin{array}{l} V_{DD} - \dfrac{K_{s\_n(sat)}\left(V_{IL} - V_{TH\_n(sat)}\right)^{\alpha\_n(sat)}}{K_{l\_p(lin)}\left(\left|-V_{DD} - V_{TH\_p(lin)}\right|\right)^{\alpha\_p(lin)/2}} - \\ - \left(\dfrac{K_{l\_n(lin)}}{K_{s\_p(sat)}} \dfrac{2}{\alpha_{n(sat)}}\right)^{-\frac{2}{\alpha_{n(sat)}+2}} \left(\left|-V_{DD} - V_{TH\_p(sat)}\right|\right)^{\frac{2\alpha_{p(sat)}}{\alpha_{n(lin)}+2}} + \\ + V_{TH\_n(lin)} \end{array} \right)^2 + \\ + \left(\begin{array}{l} \left(\dfrac{K_{l\_p(lin)}}{K_{s\_n(sat)}} \dfrac{1}{\alpha_{n(sat)}}\right)^{\frac{1}{\alpha_{n(sat)}-1}} \left(\left|-V_{DD} - V_{TH\_p(lin)}\right|\right)^{\frac{\alpha_{p(lin)}}{2\left(\alpha_{n(sat)}-1\right)}} + \\ + V_{TH\_n(sat)} - \dfrac{K_{s\_p(sat)}\left(\left|-V_{DD} - V_{TH\_p(sat)}\right|\right)^{\alpha\_p(sat)}}{K_{l\_n(lin)}\left(V_{IH} - V_{TH\_n(lin)}\right)^{\alpha\_n(lin)/2}}. \end{array} \right)^2 \end{array}} \quad (3.57)$$

3.4.4.6 Simulation Results vs. the Analytical Expression

Table 3.1 represents the extracted parameters of the α-power law model from SPICE simulations of 0.18-μm MOSFETs used in our four-transistor loadless SRAM cell. The values of α for the NMOS and PMOS transistors were obtained from fitting

Table 3.1 Alpha-power law MOSFET model parameters for 0.18 μm technology and $V_{DD} = 1.8V$ (NMOS with $W/L = 1/0.18$ μm and PMOS with $W/L = 0.5/0.18$ μm)

Model parameters	NMOS	PMOS
$\alpha\_(lin)$	1.2	1.4
$\alpha\_(sat)$	1.15	1.4
V_{D0}, V	0.98	0.98
$V_{TH(lin)}$, V, ($V_{DS} = 0.1V$)	0.49	0.522
$V_{TH(sat)}$, V, ($V_{DS} = 1.8V$)	0.489	0.488
I_{D0}, mA	0.646	0.123

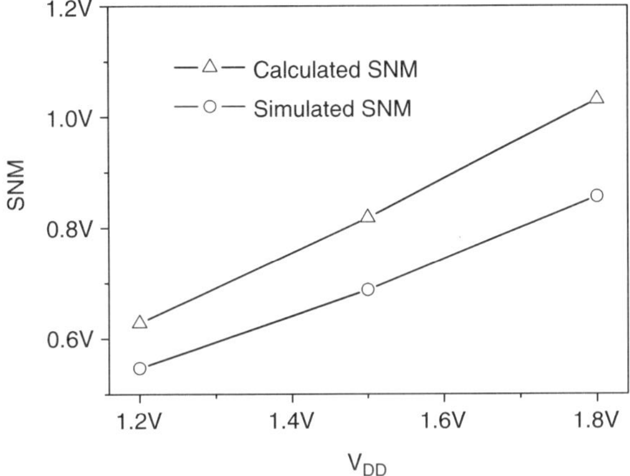

Fig. 3.14 SNM of the 4T loadless SRAM cell vs. V_{DD} ($CMOS$ 0.18 μm technology, $(W/L)_{driver} = 1/0.18$ μm, $(W/L)_{access} = 0.5/0.18$ μm). Comparison of the results using SPICE simulation and calculation using Equation 3.57

the calculated V_{GS} vs. I_D $(V_{DS} = V_{DD} = 1.8V)$ α-power model curves to the SPICE simulated ones.

The SNM definition depicted in Figure 3.13 is convenient for the analytical SNM calculation. While the SNM, defined as the diagonal or as a side of the maximum square embedded between the two inverter voltage transfer characteristics of an SRAM cell can be easily applied for graphical extraction of SNM, it cannot be derived mathematically.

Note that the proposed analytical expression (Equation 3.57), unlike the previous art [51], allows us to calculate the SNM of the loadless 4T SRAM cell accounting for the transistor parameter differences shown in Table 3.1.

Figure 3.14 presents the SNM dependence of the 4T loadless SRAM cell using the "-1" slope SNM definition on V_{DD} calculated using the proposed analytical model and simulated with HSPICE. The average error of this analytical model with respect to the simulated data for the simulated cell with $(W/L)_{driver} = 1/0.18$ μm, $(W/L)_{access} = 0.5/0.18$ μm) in CMOS 0.18 μm technology was ≃14–15% across the V_{DD} values from 1.2 to 1.8 V [52].

3.5 SRAM Cell Stability Sensitivity Factors

Figure 3.15 shows a 6T SRAM cell and its corresponding Voltage Transfer Characteristic (VTC) for a non-ideal cell.

In an ideal SRAM cell, the VTC of both halves of such a cell would be perfectly symmetrical and squares 1 and 2 between the VTC curves (Figure 3.15(b)) would be of equal size. However, in reality, process parameter spreads and the presence of defects can change the shape of the VTC curves. In the cell stability sensitivity study that follows, we define the SNM as the side of the smaller of the two squares that can be fit in the eyes of the VTC curves, as shown in Figure 3.15(b). All measurements in our analysis were taken in a *read-accessed* cell. The read-accessed cell demonstrates the worst-case SNM [51]. The saturated access transistor $Q5$ in such a cell is effectively shunting $Q3$, and $Q6$ is shunting $Q4$, which pulls up the stored low-level state and reduces the SNM as shown in Figure 3.16. Effectively, in a read-accessed cell a $CMOS$ inverter $Q2-Q4-Q6$ is turned into a $ratioed$ inverter. Consequently, the ideal logic "0" level of a CMOS inverter turns into the non-ideal "0" level, which is above the ground potential.

We investigated the SNM sensitivity to the process parameter spread (V_{TH}, L_{EFF}, and W_{eff}), the presence of non-catastrophic defects (resistive bridges and breaks), and the variation of the operating voltages of the cell (V_{DD}), (V_{BL}) and (V_{WL}). The SNM values presented are obtained using a 6T SRAM cell in 0.13 μm CMOS technology with $V_{DD} = 1.2\,V$ using special SRAM transistor models. The presented data are normalized with respect to the typical case (typical process corner, room temperature, typical voltages) according to Equation 3.58.

$$SNM_{relative} = \frac{SNM_X - SNM_{TYP}}{SNM_{TYP}} \times 100\% \quad (3.58)$$

To obtain a deeper insight into the SNM sensitivity, we simulated variations of the process parameters, operating voltages and defect resistances. The resistive defects were simulated in the range from 1 kΩ to 100 GΩ, which covers the range of

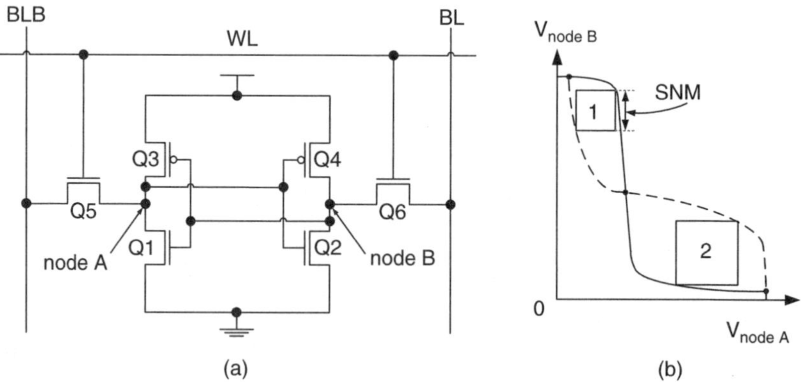

Fig. 3.15 A 6T SRAM cell (a) and its SNM definition (b)

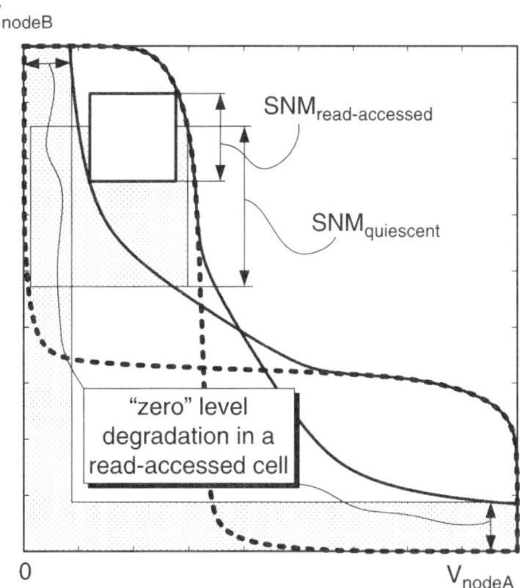

Fig. 3.16 Simulated VTCs of a 6T SRAM cell in the retention (quiescent) and in the read-accessed modes (CMOS 0.18 μm). Note that the SNM of the read-accessed cell is much smaller than that in the retention mode

possible defect resistances in an SRAM cell. In subsequent sections we use the following transistor notation (Figure 3.15(a)): $Q1$ and $Q2$ – driver or pull-down (PDN) transistors, $Q3$ and $Q4$ – load or pull-up (PUP) transistors and $Q5$ and $Q6$ – access transistors. The signal notation: BL – bit line, BLB – bit line bar, WL – word line, node A and node B – the internal nodes of an SRAM cell.

3.5.1 SRAM SNM and Process Parameter Variations

Process variations in modern CMOS DSM technologies pose an ever-growing threat to SRAM cell robustness. Threshold voltage (V_{TH}) spreads of over 10% of the typical values are not unusual anymore [41,55]. Such variations can dramatically reduce the SNM and thus, the stability of an SRAM cell, which is also demonstrated by our simulation results.

SRAM yield is correlated with the SNM spread. It is reported that, $\mu - 6\sigma$ of SNM is required to exceed $0.04 \times V_{DD}$ to reach a 90% yield on 1 MB SRAM [56]. Poor transistor matching (increased $A_{\Delta V_{TH}}$) leads to a reduction in $\mu - 6\sigma$ and increases the number of unstable SRAM cells, impacting SRAM yield. Typically, this translates into a requirement that $\text{SNM}_{min} \geq 20\%\ \text{SNM}_{typ}$.

SNM dependence on V_{TH} variations for slow, typical and fast process corners is shown in Figure 3.17.

V_{TH} was swept for only one transistor at a time while keeping the V_{TH}s of the other transistors typical. By sweeping V_{TH} of one of the transistors, we effectively introduced a mismatch between the two halves of the SRAM cell. This essentially changes the shape of the transfer characteristics (Figure 3.15(b)) and thus can

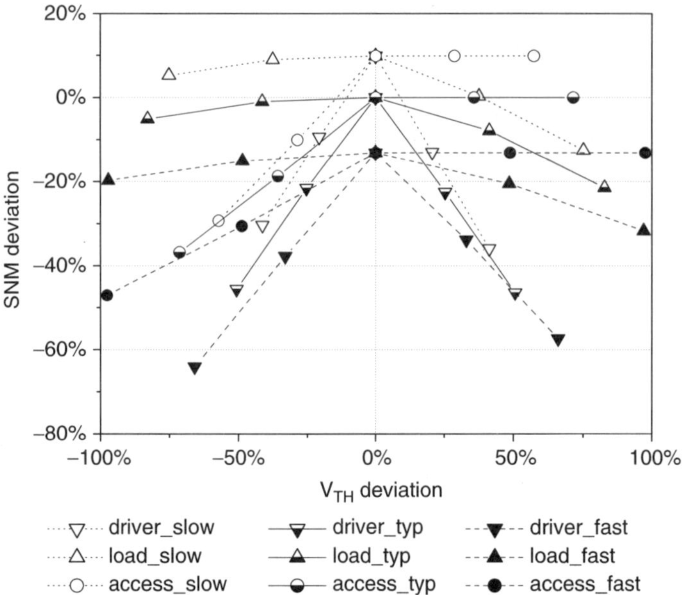

Fig. 3.17 6T SRAM cell SNM deviation vs. threshold voltage deviation of one of the transistors

adversely affect the SNM. The "0" point on the x-axis corresponds to the typical value of V_{TH} of a corresponding transistor in the corresponding process corner.

V_{TH} variation of the driver transistor had the largest impact on the VTC shape and thus SNM due to its larger W/L ratio compared to other transistors in SRAM cell. Decreases in the V_{TH} of the access transistor also has a strong negative impact on the SNM. Since the measurements were taken in a read-accessed SRAM cell, the access transistors are effectively connected in parallel with the load transistors. Thus, reducing the V_{TH} of the access transistor compromises the low level stored in the cell, which in turn reduces the SNM. On the other hand, the V_{TH} variation of the PMOS load transistor has the least impact on the SNM due to its weaker drive and typically smaller W/L ratio.

Note that the SNM deviation is zero if the V_{TH} deviation of all transistors is zero (symmetrical cell), except only for the case of increasing V_{TH} of the access transistor, which does not affect the SNM as its shunting action on the load transistor decreases.

If more than one V_{TH} is affected at a time, the SNM degradation can be stronger. Figure 3.18 presents several cases of the SNM vs. V_{TH} dependencies when V_{TH} of more than one transistor in the SRAM cell is not at its typical value (typical process corner). For instance, $Q2\_(Q1 = -25\%)$ in Figure 3.18 represents the dependence of the SNM on the V_{TH} of $Q3$ provided that V_{TH} of $Q1$ is below its typical value by 25%. This dependence has its maximum at the point where $V_{TH\_Q1} = V_{TH\_Q2} = -25\%$ (i.e., where the cell is symmetrical). SNM vs. V_{TH} of $Q5\_(Q1 = -25\%, Q2 = +25\%, Q3 = +40\%$, and $Q4 = -40\%)$ represents one of the worst cases of the SNM degradation due to the asymmetry of V_{TH} of the cell's transistors.

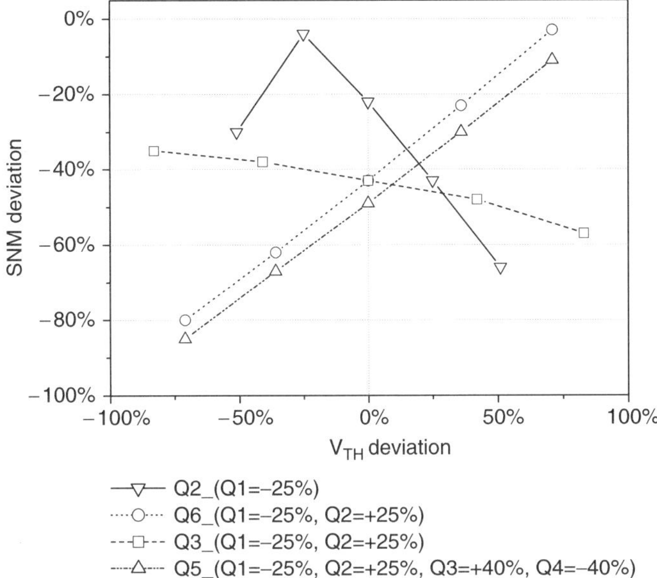

Fig. 3.18 SRAM cell SNM vs. threshold voltage deviation of more than one transistor

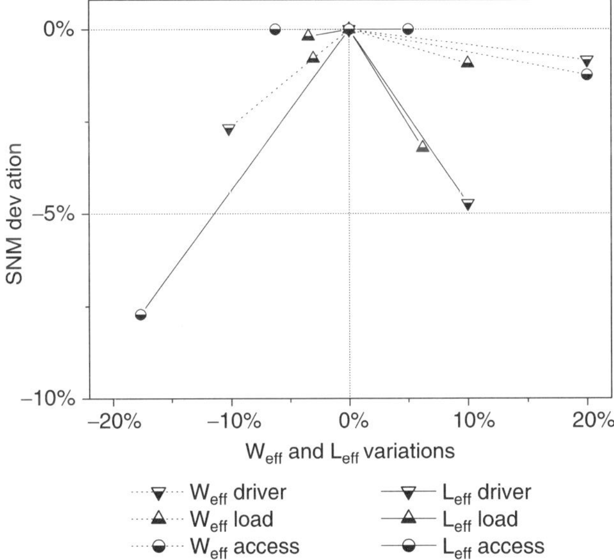

Fig. 3.19 SRAM cell SNM deviation vs. L_{EFF} and W_{EFF}

SRAM cell SNM dependence on L_{EFF} or W_{EFF} variations in a single transistor under typical conditions is shown in Figure 3.19. The SNM decrease is insignificant when the transistor's effective length and width variation remains within 20% of the typical values. Regardless of the direction of the transistor geometry variation, SNM is maximum at the typical (symmetrical) transistor sizes. This is because the variation of the transistor geometry in only one of the halves of the SRAM cell causes mismatch, which leads to the reduction of the SNM. Figure 3.19 shows that for a weaker driver transistor (smaller W/L ratio) the SNM decreases, whereas a weaker access transistor improves the SNM. Deviations in W/L of the load transistor just slightly degrade the SNM.

Figures 3.17, 3.18 and 3.19 show that the SNM of an SRAM cell is maximum if the effective driving strength of both halves ($Q1-Q3-Q5$ and $Q2-Q4-Q6$) of the cell is symmetrical with respect to their threshold voltages and W/L ratios.

As the technology is scaled into the decananometer region, V_{TH} spread becomes one of the main contributors to wider distribution of the SNM figures even in a defect-free chip. The granularity of the electric charge and the atomic nature of matter introduce substantial variation in number and position of dopant atoms in the transistor channel, t_{ox} becomes equivalent to several atomic layers with one to two atomic layers roughness [57]. For sub-100 nm CMOS SRAMs, 6σ deviations of the SNM only due to intrinsic device fluctuations are projected to exceed the nominal SNM [41]. Randomness of channel dopant distribution will become a major source for the SNM reduction. To maintain reasonable SNM and yield in future bulk CMOS SRAMs, cell ratios may have to be increased from the conventional $r=\frac{(W/L)_{driver}}{(W/L)_{access}}=2$ towards the higher ratios [35], which counter-balances the scaling advantages of deep sub-micron technologies with respect to the area of the embedded SRAM cores.

3.5.2 SRAM SNM and Non-catastrophic Defects

The defects that cause a functional failure of an SRAM cell can be defined as *catastrophic* defects. Most catastrophic defects in SRAM cells tend to be easily detected by the regular march tests. If a defect in the cell does not cause a functional fault in the normal test and operating conditions, such a defect is defined as a *non-catastrophic*. SRAM cells with non-catastrophic defects can have non-zero SNM and escape standard functional tests. Besides degrading the cell stability, if left undetected, the non-catastrophic defects can pose long-term reliability issues.

3.5.2.1 SNM vs. Non-catastrophic Breaks and Bridges

In order to investigate SRAM SNM degradation as a function of the resistance of non-catastrophic defects, we used the Carafe Inductive Fault Analysis (IFA) [58] tool on a cell similar to the one shown in Figure 4.3. Carafe introduces *resistive* defects in the layout [59] of the cell. It works by widening and shrinking the layout geometries and finding possible intersections of conductors in various process planes to determine how a spot defect of a certain size can affect the layout. Based on the layout sensitivity analysis, Carafe generates a fault list. The most probable faults were simulated to extract the SNM values. We modeled the obtained faults as parallel and series resistors to represent bridges and breaks (opens) in the cell, respectively. Since an SRAM cell has a symmetrical structure, certain defects can appear in either half of the cell doubling the probability of the corresponding defects and promoting these defects on the list. Defects with resistance from 1 kΩ to 50 GΩ for both the breaks and the bridges were injected one at a time, while keeping all other conditions typical.

Figures 3.20 and 3.21 are showing the SNM as a function of the most likely breaks (resistive opens) and bridges (resistive shorts), respectively. As it is evident from Figures 3.20 and 3.21, increasing the resistance of the bridges and breaks in an SRAM cell has an opposite effect on the cell's SNM. When the resistance is increasing, most resistive opens (breaks) linearly degrade the SNM. Above certain resistance values, they cause the SNM to become a zero. Resistive open defects are likely to appear in place of poor or absent contacts, vias or silicide [1, 2]. Note that different breaks can have different impact on the SNM. A break in the local ground contact of SRAM cell has the strongest negative impact on the SNM. A break in the drain of a driver transistor (D_{driver}) also causes a severe reduction of the SNM. A break in a cell's (local) V_{DD} or in the drain of a single load transistor (D_{load}), which can cause Data Retention Faults (DTFs), has a medium impact on the SNM. In contrast, resistive breaks in transistor gates do not cause a noticeable SNM degradation unless the break resistance exceeds approximately 1 GΩ.

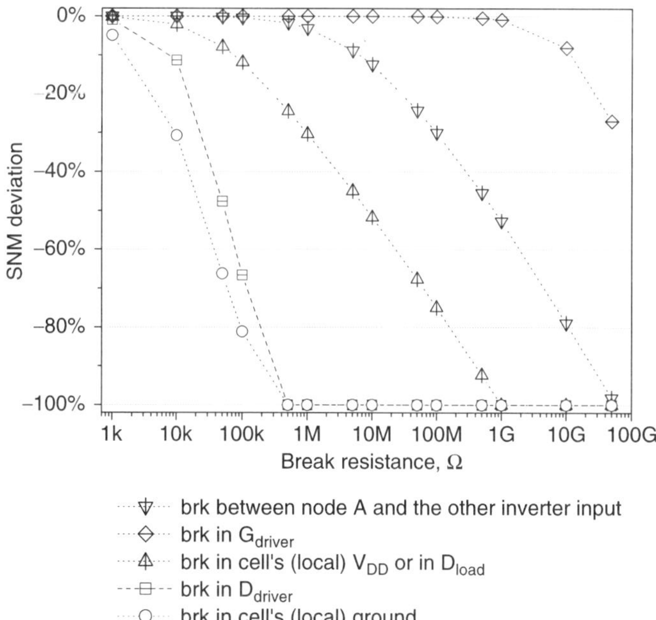

Fig. 3.20 SRAM cell SNM deviation vs. break (resistive open) resistance

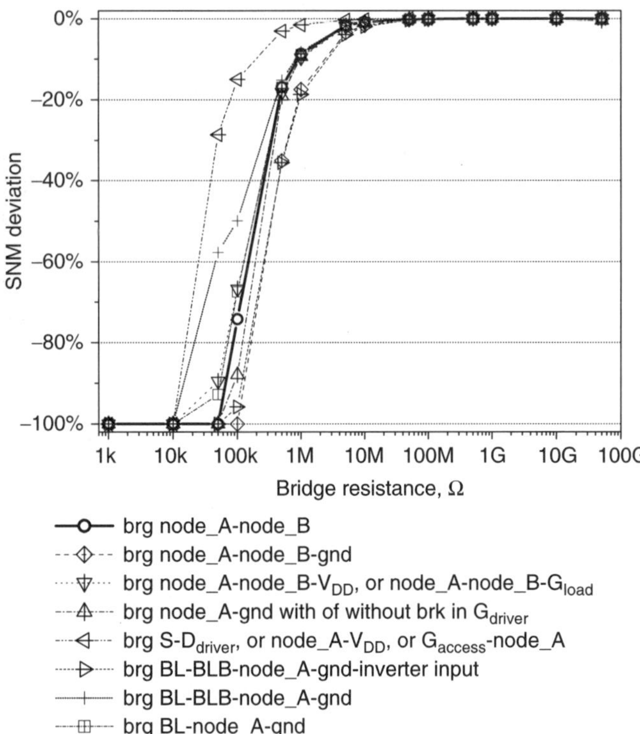

Fig. 3.21 SRAM cell SNM deviation vs. bridge (resistive short) resistance

Figure 3.21 shows that in contrast to the SNM dependence on the break defect resistance, a reduction in the resistance of most bridges causes a very similar degree of SNM degradation. Bridges with resistance below 10–100 kΩ reduce the SNM to be near zero and cause catastrophic functional failures, which are easily detected by the regular march tests. The SNM increases almost linearly for most of the bridges having resistances between 100 kΩ and 1 MΩ. Bridge defects with a resistance of more than 10 MΩ have virtually no impact on the SNM. Due to the cross-coupled layout of the SRAM cell under consideration, the most likely resistive bridge in this SRAM cell is the bridge between the two internal nodes. In Figure 3.21, the SNM dependence on this bridge (brg node_A-node_B) resistance is shown in a bold solid line.

3.5.3 SRAM SNM and Operating Voltages Variation

Variations in the operating voltages, such as the supply (V_{DD}), the bit line (V_{BL}) or the word line (V_{WL}) voltages, strongly impact the stability of an SRAM cell. The SNM dependencies on the operating voltages can reveal valuable information on the stability margins of a particular cell. The worst-case SNM is typically observed for the fast process corner and high temperature, while the best case SNM occurs for the slow process corner and low temperature. The results for all other temperature/process corner combinations fall in between the best and the worst cases described above. As a demonstration of the typical behavior of an SRAM cell under varying operating voltages, we will present the dependencies of the cell's SNM on V_{BL}, V_{DD} and V_{WL}. Figures 3.22, 3.23 and 3.24 show the SNM as a function of the variation of a single operating voltage.

Figure 3.22 depicts the SNM dependence on the bit line voltage V_{BL} while the V_{DD}, V_{WL}, and V_{BLB} are at the typical 1.2 V. The situation when one of the bit lines is driven from V_{DD} to the ground corresponds to a "write" operation. Overwriting the data stored in an SRAM cell becomes possible when the SNM becomes zero. As is apparent from Figure 3.22, the SNM in the typical process corner becomes a zero at $V_{BL} < 0.3$ V. Note that the SNM does not decrease immediately once V_{BL} starts decreasing. The reduction of V_{BL} begins to reduce the SNM once $|V_{BL} - V_{WL}| > V_{TH_{access}}$ and the access transistor turns on. Since transistors in the slow process corner have higher V_{TH}s, with the reduction of V_{BL} the cell SNM stays constant for a longer period than its counterparts from the typical and the fast process corners.

A write operation is possible in the region where the bit line voltage is at or below a voltage point where the SNM becomes a zero. This voltage region is sometimes called the *write margin*.

The write margin is an important design parameter as it defines the "writeability" of the cell. A balance between the cell stability (SNM), the writeability, the cell area and the access speed must be found, such that the cell has a sufficient stability margin while it still can be reliably written into.

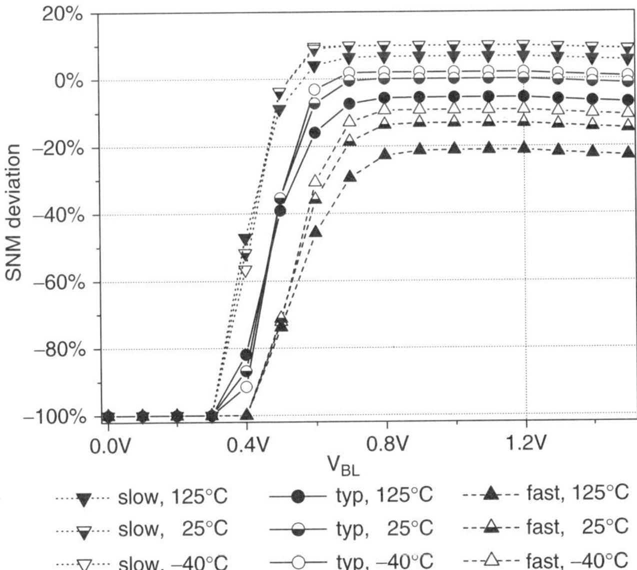

Fig. 3.22 SRAM cell SNM deviation vs. bit line voltage V_{BL}

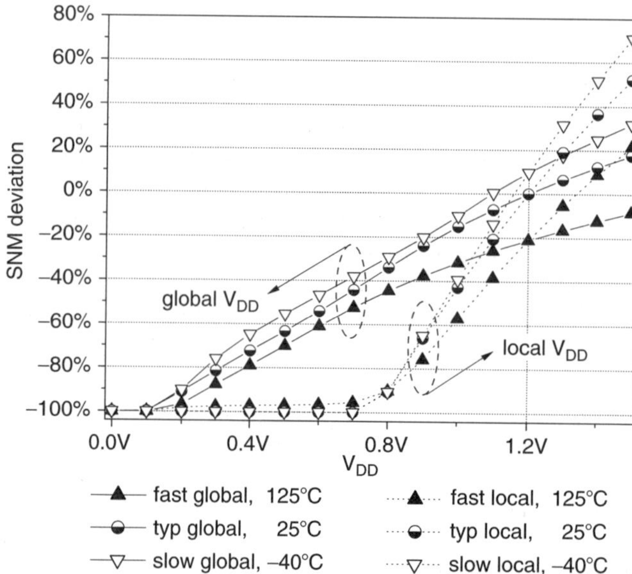

Fig. 3.23 SRAM cell SNM deviation vs. the power supply voltage V_{DD}

Figure 3.23 shows SNM deviation as a function of the global and local V_{DD}. By global V_{DD} variation, we mean the situation when V_{WL}, V_{BL}, V_{BLB}, and SRAM cell supply voltages vary all at the same time. This could model a battery discharge in a mobile device or a gradual scaling down V_{DD}/clock frequency during the low activity periods to achieve power savings. By local V_{DD}, we imply only the variation of SRAM cell supply voltage, while V_{WL}, V_{BL} and V_{BLB} are all at the typical values. Local V_{DD} variation mimics a faulty via in the supply voltage grid of SRAM cell.

SNM shows strong virtually linear proportional dependence on both the local and the global V_{DD} variations. However, the SNM dependence on the local variation is stronger since in this case the word line and the global bit lines are at full V_{DD}. This causes the access transistor to shunt the pull-up transistors more strongly and degrade the low state of the cell. Note that by raising the local V_{DD}, we can observe a significant increase of the SNM because the drive of the access transistors of the read-accessed cell is becoming weaker while the power supply of the cross-coupled inverters rises. Due to lower V_{TH_n}, the fast process corner cases exhibit lower SNM values as opposed to the slow corner cases.

Figure 3.24 shows the dependence of the SNM on the word line voltage, while all other operating voltages are at their typical values. While V_{WL} is below the V_{TH} of the access transistor, the cell can be considered to be in the standby or quiescent mode and its SNM is not affected. Once $V_{WL} > V_{TH}$ of the access transistor, the SNM starts to deteriorate as the access transistor starts to shunt the "off" load transistor and pull the node storing a logic zero up. Note that if the word line exceeds V_{DD}, the SNM continues to deteriorate as the shunting action of the access transistor strengthens.

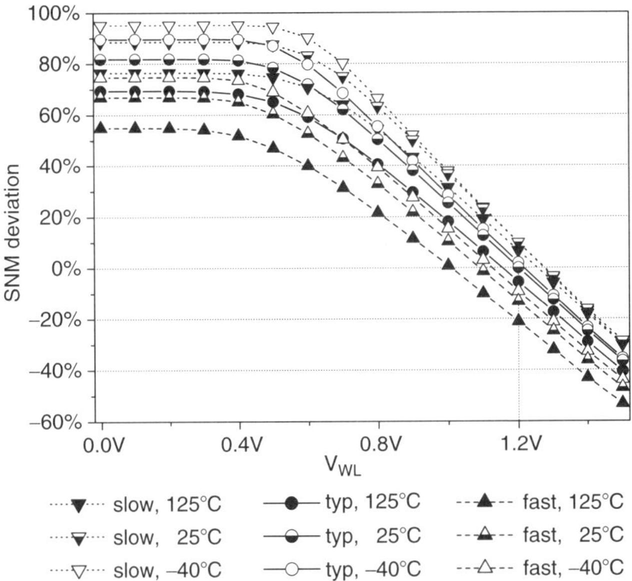

Fig. 3.24 SRAM cell SNM deviation vs. word line voltage V_{WL}

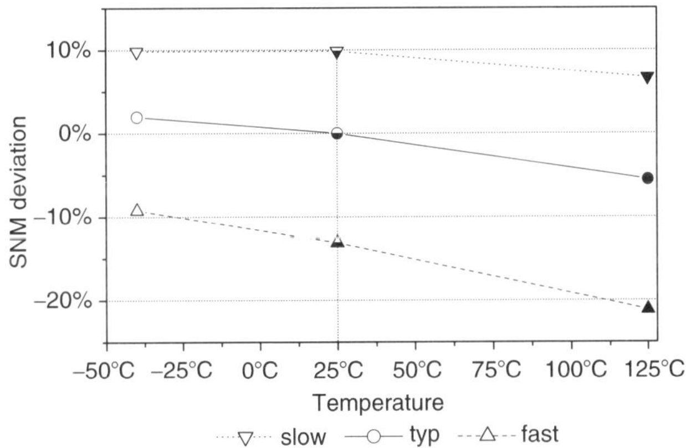

Fig. 3.25 SRAM cell SNM deviation vs. temperature T

Among other parameters, the operating temperature has an insignificant effect on the SNM. If all voltages are kept at their typical values and the temperature is varied from $-40°C$ to $125°C$, the SNM demonstrates rather weak temperature dependence from $\simeq 5\%$ for the slow process corner to $\simeq 12\%$ for the fast process corner, as it is apparent from Figure 3.25. The SNM tends to decrease at the elevated temperatures.

In the previous paragraphs, we described the impact of a single process parameter variation on the SNM. However, in real life, several process parameters may change simultaneously and randomly. In case more than one process parameter (especially V_{TH}) departs from its typical value, the impact on the SNM can be dramatic, often reducing the SNM to a very low value. Such SRAM cells are prone to stability faults, which may escape standard tests. Stability faults can thus potentially manifest themselves as long-term reliability problems that must be addressed by thorough testing and repair.

3.6 SRAM Cell Stability Fault Model

An accurate and consistent representation of unstable behavior of an SRAM cell is essential for SNM modeling and evaluation of circuit techniques for SRAM cell stability test. Since the SNM is a measure of SRAM cell stability, its degraded value results in a cell stability fault that is parametric in nature. This fault can manifest itself under certain conditions by compromising the stored data integrity. A parametric stability fault model can be used for investigation and comparison of the effectiveness of various test algorithms and DFT techniques as well as for the stability characterization of SRAM designs. Such a stability fault model [6] is presented below.

Let us consider the dependence of the SNM on the resistance between node *A* and node *B* as illustrated in Figure 3.26, and also in a thicker solid line in Figure 3.21. For very large resistance values ($>$10 MΩ), the cell SNM is not affected. For the resistance range between 50 kΩ and 1 MΩ, the SNM is reduced linearly. The SNM becomes zero and causes catastrophic failure for the resistance values below 50 kΩ. Depending on the parameters of a given SRAM cell, one can choose a particular resistance value in order to realize a weak cell with a pre-determined SNM. For instance, to obtain a cell with a half of the typical SNM, a 200 kΩ resistor must be used. A bold line in Figure 3.26 represents a possible target range for weak SRAM cell detection.

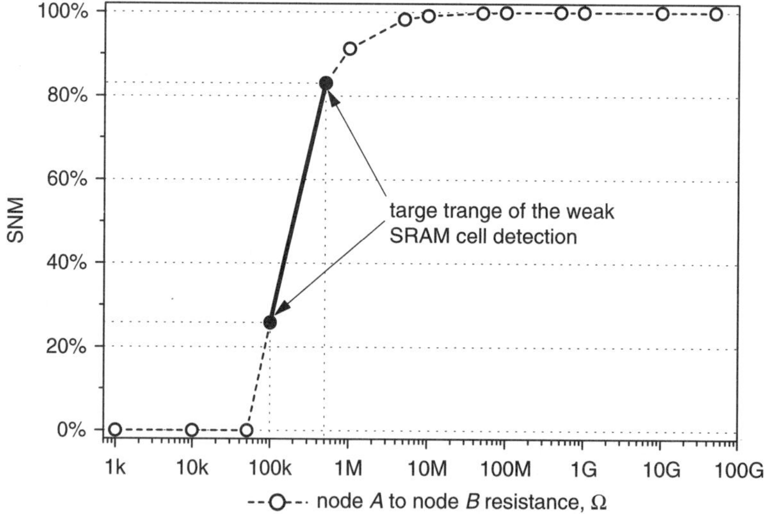

Fig. 3.26 A possible target range of the SNM modeled by a given range of a resistor between node A and node B (0.13 μm technology) [6]

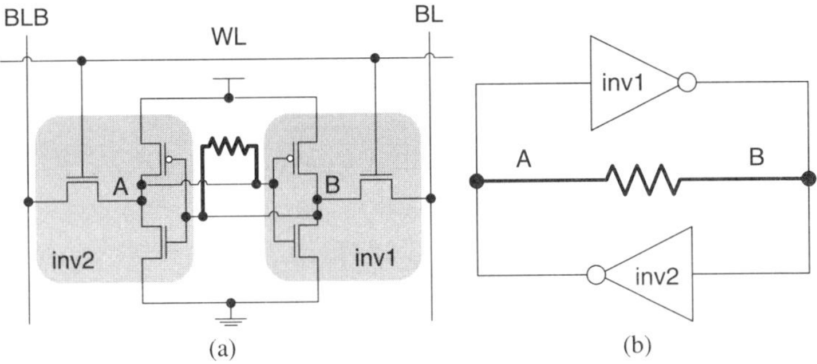

Fig. 3.27 Weak Cell Fault Model (a) and its equivalent circuit (b) [6]

The resistor between node *A* and node *B* represents the proposed *weak cell* or *cell stability* fault model [6], which is illustrated in Figure 3.27(a). This SRAM cell has the worst-case SNM in the read-access mode when both the bit lines are precharged and the word line is activated [51]. Each half of a read-accessed SRAM cell can be represented as an equivalent inverter, as shown in Figure 3.27(b). As we can see from Figure 3.27, a node-to-node resistive defect represents a negative feedback for the equivalent inverters comprising the SRAM cell. The corresponding reduction of the inverter gains and hence, the amount of the negative feedback in the cross-coupled inverters, is symmetrical and can be used to control the SNM. In a simulation environment, a cell with a resistor of a specified value between node *A* and node *B* can imitate a weak cell with a specified SNM value. The degree of the "weakness" is controlled by the value of the resistor. Provided other conditions are equal, the "weakened" cell has equal SNMs and symmetrical response for both high-to-low and low-to-high internal node voltage transitions. Thus, it represents a simple, symmetric and realistic stability fault model for simulation of parametric stability faults in SRAMs.

Intentionally inserting weak cells with the desired target SNMs into an SRAM array allows us to verify and fine-tune test techniques for parametric stability fault (weak cell) detection in the simulation environment. Having a simulation setup with a set of weakened cells with varying degrees of weakness (SNM) allowed us to evaluate various cell stability DFT techniques and algorithms. Moreover, such a setup can also be instrumental in the stability characterization and debugging stages of SRAM development.

3.7 SRAM Cell Stability Detection Concept

To provide better insight into the principles behind the detection of weak SRAM cells, we developed an SRAM cell stability detection concept. To illustrate the concept consider the voltage transfer characteristics of a good SRAM cell (solid lines) and a weak SRAM cell (dashed lines) presented in Figure 3.28. The axis in Figure 3.28 represent node A and node B voltages, which in turn, are proportional to the bit line voltages V_{BL} and V_{BLB}.

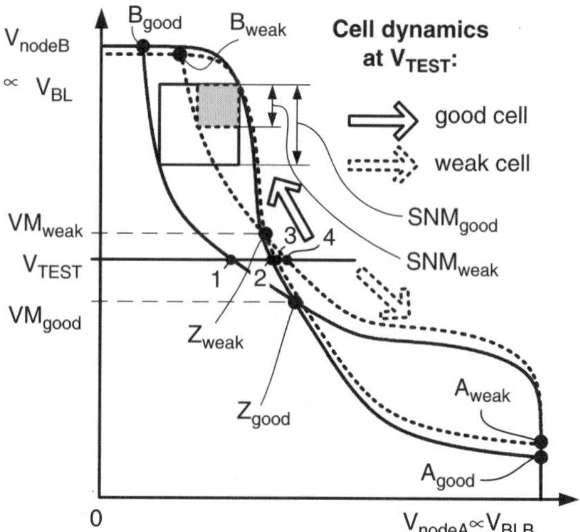

Fig. 3.28 Choice of V_{TEST} with respect to the VTCs of a typical and a weak SRAM cell

Normally, the real VTC of an SRAM cell is asymmetrical due to inevitable mismatches or defects. The SNM of the cell is proportional to the degree of asymmetry of the VTCs $((SNM_{weak} < SNM_{good}))$. Fluctuations in V_{TH} and L_{EFF}, the presence of defects, and poorly formed contacts and vias can weaken the driving strength of one of the cell inverters. This shifts the meta-stability point of the cell. Without loss of generality, suppose that a given cell's VTC is skewed so that its metastable point VM_{weak} is closer to node B. Since the metastable point is not equidistant from the the node potentials, the affected data node will be more vulnerable to disturbances than the other node. Any noise disturbance exceeding the metastable point of the cell will cause such a cell to flip states. In other words, if a data node of an SRAM cell is driven to the level of VM, then a small voltage increment will flip the cell towards the direction of this increment.

VM_{good} and VM_{weak} represent node B voltages corresponding to the metastable points Z_{good} (Z_{weak}) of the good (weak) cell, respectively. Note that the metastable point of a weak cell is different from the metastable point of a good cell. The difference is proportional to the degree of asymmetry introduced into the weak cell VTC by a defect or a mismatch. This property of cell's VTCs can be exploited in the cell stability test techniques.

Let us assume that node B of an SRAM cell stores state "1" and that the bit lines are pre-charged to a known value (e.g. V_{DD}). Now assume that by a certain manipulation, V_{node_B} is reduced from a stable state B_{good} (or B_{weak} for a weak cell) to a certain test voltage V_{TEST}. Voltage level V_{TEST} intersects the good cell's and the weak cell's transfer characteristics at points "1" and "2" and at points "3" and "4" respectively, as shown in Figure 3.28. It is apparent from Figure 3.28 that the weak cell will flip its state if $(V_{DD} - V_{TEST}) < (V_{DD} - VM_{weak})$ in the direction of the dotted arrow. In other words, the weak cell will flip if its node B is driven below V_{TEST}.

If $V_{TEST} > VM_{weak}$, the regenerative property of a weak cell will restore the stable state and the weak cell will not flip. This situation is similar to a non-destructive read operation of the cell under test with incompletely precharged bit lines. If $V_{TEST} < VM_{good}$, even the good cells will flip. This situation is similar to a normal write operation on the cell under test. V_{TEST} is in the target range if $VM_{good} < V_{TEST} < VM_{weak}$. This is the selectivity condition of weak cell detection during the stability test.

Upon removal of the test stimulus V_{TEST} node B of the good cell will retain its state "1", while node B of the weak cell will flip to state "0". The arrows in Figure 3.28 show the direction of the cell dynamics at V_{TEST}. After carrying out a cell stability test, all the cells, which flip after application of the node voltage above V_{TEST} are deemed "weak". The rest of the cells is assumed to have acceptable stability.

Test voltage V_{TEST} can have a fixed or a variable value. A fixed V_{TEST} allows for a single pass/fail threshold, whereas being able to vary the V_{TEST} value, one can test for a given degree of cell weakness and shift the pass/fail threshold according to the target quality requirements. Various cell stability test techniques presented in Chapter 5 can be classified based on this and other principles.

Variable V_{TEST} that is typically achieved by digitally programming enables better process variation tracking and allows to adjust the pass/fail threshold *during the test*. The importance of the ability to program the pass/fail threshold can be understood by inspecting Figure 3.29.

Figure 3.29(a) shows the case when V_{TEST} is outside of the target range such that $V_{TEST} > VM_{weak}$. Applying V_{TEST} which is higher than the target range will not exert sufficient stress to flip the weak cells and the weak cells will escape the test undetected. The case in Figure 3.29(b) shows V_{TEST} positioned correctly between

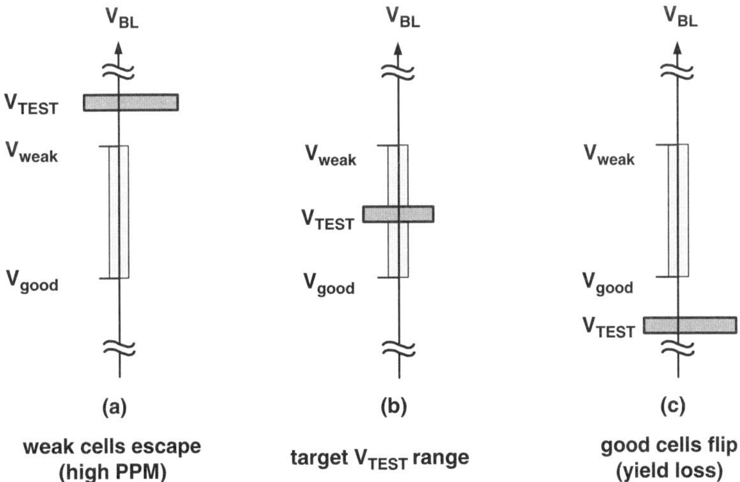

Fig. 3.29 Programming V_{TEST} to set a correct pass/fail test threshold: (a) V_{TEST} is set too high, which causes the weak cells to escape (high PPM), (b) V_{TEST} is set correctly, (c) V_{TEST} is set too low, which causes the good cells to flip (yield loss)

VM_{good} and VM_{weak}. And finally, Figure 3.29(c) shows the case when V_{TEST} is outside of the target range such that $V_{TEST} < VM_{weak}$. Applying V_{TEST} which is lower than the target range will exsert too much stress causing even the good cells to flip leading to the yield loss. Cases (a) and (c) are likely to happen in case of severe process variations or using process splits during the product yield ramp-up. Ability to program the pass/fail threshold in fine steps provides for adjusting the pass/fail threshold on the fly without the need to redesign the DFT through multiple design iterations.

3.8 March Tests and Stability Fault Detection in SRAMs

Functional march tests, reviewed in Section 4.3.1, remain the main SRAM testing method [60]. By optimizing a march test suite to a particular SRAM architecture and technology, the fault coverage of the "hard" faults was sufficient in many cases to ensure acceptable PPM levels that are approaching single-digit numbers [61]. However, with technology scaling, the growing probability of stability faults [6] and dynamic faults [62,63] can cause a higher PPM levels, which is often unacceptable.

In this section we discuss the capabilities of March 11N and hammer tests of detecting Stability Faults and Dynamic Faults in SRAM arrays.

3.8.1 March 11N

We evaluated the detection capabilities of an 11N march test on an 130 nm SRAM array consisting of 4 x 4 cells. One of the cells in the array was intentionally weakened by reducing its SNM [64]. The SNM reduction to a desired value was achieved by varying the resistance of a resistor between the two data nodes. This resistor represents the stability fault model presented in Section 3.6.

Table 3.2 summarizes the detection capabilities of the 11N March test for different process corners (cycle time 2.4 ns, 130 nm technology [64]). The reduced SNM values with respect to the typical SNM for the corresponding process corners shown in Table 3.2 correspond to the resistances of R placed between the two data nodes A and B of an SRAM cell.

Table 3.2 Summary of March 11N effectiveness in detecting weak SRAM cells (cycle time 2.4 ns, 130 nm technology [64]), where R is the resistor between the two data nodes of an SRAM cell representing the stability fault model presented in Section 3.6

	Slow corner	Typical corner	Fast corner
$R_{not\_detected}$	100 kΩ	50 kΩ	30 kΩ
$SNM_{not\_detected}$	27%	4%	2%
$R_{detected}$	90 kΩ	47 kΩ	25 kΩ
$SNM_{detected}$	21%	3%	1%

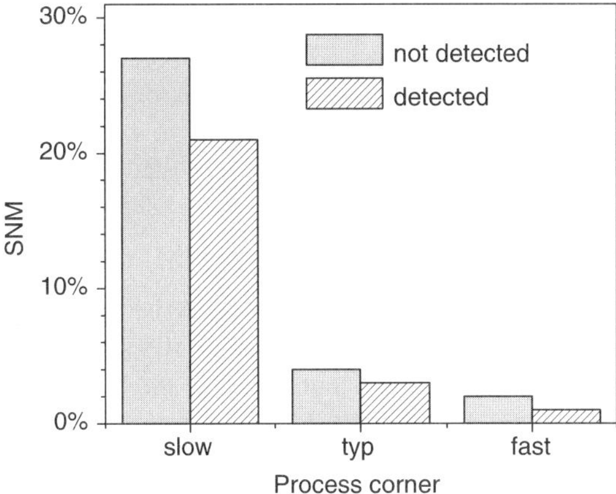

Fig. 3.30 Relative SNM values of a weakened SRAM cell that can be detected by a March 11N test at various process corners (130 nm technology) [64]

Application of the March 11N test for the slow process corner is ineffective for the detection of a weak cell with a $R_{node\ A-node\ B} = 100\,\mathrm{k\Omega}$. However, a cell with $R_{node\ A-node\ B}$ of $90\,\mathrm{k\Omega}$ caused a destructive read, which was detected by the March 11N.

Table 3.2 shows that while the SNM value that can be detected by March 11N test in the slow process corner is 21% of the the full typical SNM value, March 11N applied to an SRAM cell with typical process parameters can only detect the SNM drop below 3%. In the fast process corner the detected SNM value drops even more to just 1%.

Figure 3.30 graphically represents the decline in weak cell detection capability with the process corner variation from slow to fast. Apparently, March 11N test is incapable of reliable detection of the stability faults even for the typical SRAM cell fabricated with the typical process target parameters. An SRAM cell with just a 3% SNM margin that can be detected by March 11N at typical process corner is highly vulnerable to disturbances and its stability is clearly insufficient for any practical purposes.

3.8.2 Hammer Test

Most of the SRAM test methods targeted functional fault models limited to static faults. Static faults are faults that can be sensitized by performing one operation. However, dynamic faults have been reported recently based on defect injection and fault simulation of the industrial SRAMs [62]. Dynamic faults may have ton be sensitized by more than one operation sequentially. They can be caused by the same reasons as the stability faults, i.e. mostly by mismatches, defects and adverse environmental conditions. Since only a defective cell can exhibit a dynamic fault, in the context of this book Dynamic Faults and Stability Faults can be treated as equivalent.

The hammer test is classified as a Repetitive Test [61], i.e., a test that repetitively applies multiple write or read operation to a single cell. Repeating the test makes partial, hard-to-detect fault effects become full fault effects. Equation 4.3 presents a hammer test with test length of 49N.

$$\Updownarrow (w0) \Uparrow (r0, 10*w1, r1) \Downarrow (r1, 10*w0, r0)$$
$$\Uparrow (r0, 10*w1, r1) \Downarrow (r1, 10*w0, r0) \tag{3.59}$$

The main feature of this hammer test is that the write operation is performed on the same cell ten times successively. It is denoted as $10*w1$ or $10*w0$ for a "1" and a "0", respectively. The address increment order for the hammer test was assumed "fast column" as opposed to the interlaced "fast column" and "fast row" of the 11N March test in the previous section.

However, the detection capabilities of the hammer test with respect to stability faults are also shown to be inadequate. Similar to the 11N March, the hammer test can detect only unacceptably low SNM values. For instance, the largest detected $R_{node\ A-node\ B}$ is $60\,\mathrm{k}\Omega$, which translates into the SNM value of just about 4% of the typical [64].

Many variations of the hammer test are possible depending on the number of times each read or write operations occur in each element of a hammer test [61]. A modified version of the hammer test having $3*w$ and $2*r$ operations in all but the first test elements which was also evaluated for its stability detection capabilities is presented in Equation 3.60:

$$\Updownarrow (w0) \Uparrow (2*r0, 3*w1, 2*r1) \Uparrow (2*r1, 3*w0, 2*r0)$$
$$\Downarrow (2*r0, 3*w1, 2*r1) \Downarrow (2*r0, 3*w1, 2*r1) \tag{3.60}$$

However, the modification of the hammer test did improve its fault detection capability. The SNM detected by the modified hammer was also just 4%, which is extremely low and unacceptable for the reliable detection of stability faults in SRAM cells.

3.8.3 Coupling Fault Detection

In this work we established the detection capabilities and test potential of March tests in the presence of resistive and capacitive coupling between the aggressor and the victim cell nodes as well as the bit lines of the neighboring columns. Coupling between the neighboring rows was not considered due to power busses running between the rows on the same metal layer. A low-resistance bridge spanning across such busses would likely cause a stuck-at fault in the affected cells and would therefore be easily detected by the standard test methods.

Applying the principles of the IFA, we analyzed the layout of the cell under test and defined the two most probable bridging resistive faults between two neighboring

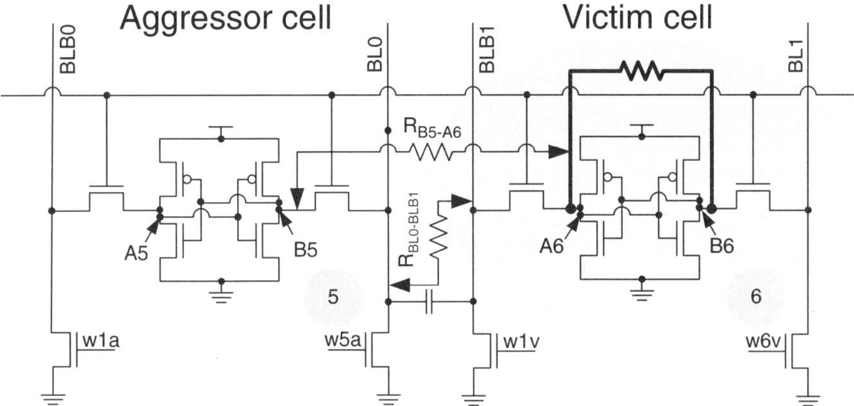

Fig. 3.31 Coupling resistive bridge defects (R_{B5-A6} and $R_{BL0-BLB1}$) and inter-bit-line capacitance between the bit lines of the aggressor and the victim cells

columns that are shown in Figure 3.31 [64]. Resistors R_{B5-A6} and $R_{BL0-BLB1}$ represent a resistive bridge between node B of the aggressor cell #5 (B5) and node A of the victim cell #6 (A6). The resistor between nodes A6 and B6 of the victim cell represents the Cell Stability Fault Model described in Section 3.6.

We developed a short march test consisting of two march elements for the test bench shown in Figure 3.31:

$$(w1v, rv, w1a, rv); (w0v, rv, w0a, rv). \tag{3.61}$$

where $w1v$ and $w0v$ denote write "1" and "0" operations on the victim cell; $w1a$ and $w0a$ denote write "1" and "0" operations on the aggressor cell; and rv and ra denote read operations on the victim and the aggressor cells respectively.

Each test element starts out by writing the background value to the weak victim cell #6 followed by a read operation on the same cell to establish a reference value. Next, the same data background is written to the aggressor cell #5. Since the aggressor and the victim cells are in adjacent columns, the data backgrounds written in the cells effectively result in opposite values stored on node A5 and node B6. If a bridging defect with sufficiently low resistance R_{B5-A6} or $R_{BL0-BLB1}$ is present between the columns, the victim cell can be inadvertently overwritten by the aggressor cell. The following read operation will determine whether the victim cell has changed states. The second march element is the inverse of the first and is needed for testing the victim cell's susceptibility to the aggressor cell with the inverse data background.

The second read operation in each of the march elements is redundant and was introduced for observation convenience. It can be removed without sacrificing the detection capability. The optimized test sequence is shown below:

$$(w1v, w1a, rv); (w0v, w0a, rv) \tag{3.62}$$

In addition to the conditions specified in Equation 3.62, the successful detection of a weak cell depends on the resistance values of R_{B5-A6} or $R_{BL0-BLB1}$ and the severity of the SNM degradation in the victim cell. All other conditions being equal, aggression on the node of the victim cell that stores a "1" overwrites the victim cell more easily. This is understood keeping in mind the design principles of 6T SRAM cells. The PMOS pull-up transistor is normally designed to be 1.5–2 times weaker than the NMOS driver transistor. Multiplying the size ratio by μ_n/μ_p shows that the current ratio $I_{pull-down}/I_{pull-up}$ can be significant. Therefore, if node "1" of the cell with compromised stability is coupled to the aggressor via a sufficiently low resistance of a bridging defect, it is more likely to be overwritten. Node "0" of the victim cell is vulnerable to the inadvertent overwriting when coupled by an aggressor with a smaller bridging defect resistance.

Experimenting with resistance values of R_{B5-A6}, $R_{BL0-BLB1}$ and R_{A6-B6} showed that the developed test sequence can detect resistive coupling faults below 10 kΩs. Bridging defects of this resistance range are usually targeted by other memory tests as well.

The test is more sensitive to R_{B5-A6} than to $R_{BL0-BLB1}$, i.e. it can detect somewhat higher values of R_{B5-A6}. The aggressor and the victim cells share the same word line. If a $R_{BL0-BLB1}$ bridge exists, then during the writing of a "0" to the aggressor cell, the precharge in the aggressor column is turned off and BL0 is discharged to the ground. The ground potential is then coupled to the victim cell via the path consisting of the access transistor of the aggressor cell, $R_{BL0-BLB1}$, and the access transistor of the victim cell. Since the precharge in the victim column is on, BLB1 is held at V_{DD} by a strong precharge transistor which is counteracting the disturbance. Once the access cycle is over, the aggressor and the victim cells are disconnected from their bit lines and $R_{BL0-BLB1}$ has no effect on their contents. R_{B5-A6} on the other hand, couples the victim and the aggressor cells all the time regardless of whether this particular row is active or not. A write "0" operation to node B5 in this case can destroy a "1" stored on node A6 more easily.

Detection of a coupling defect R_{B5-A6} in the case when the aggressor node B5 is written a "1" and the victim node A6 stores a "0" is very unlikely unless the victim cell's pull-down NMOS is seriously damaged. In fact, it can even cause the situation, when a write access to the victim cell destroys the data in the aggressor cell, i.e., the aggressor and the victim cells would change places. A damaged NMOS pull-down transistor reduces the I_{read} of the cell and is easier to detect as it may fall out of the read timing specification. Such defects can be easily detected by simple functional memory tests.

Capacitive coupling between the bit lines of the neighboring columns did not play a notable role. In the considered layout the bit lines are divided by a ground wire on the same metal layer. Even in a layout without a ground wire between the columns, the capacitance between the neighboring columns is unlikely to cause a coupling fault in many SRAM architectures. Typically, bits of the same significance belonging to different bytes/words are laid out in adjacent columns and share a common SA through the column MUX. When write accessed, only one of the neighboring bit lines can be driven at a time from the precharge potential (V_{DD}) to the ground.

At the same time, the neighboring bit line is held at V_{DD} by a strong precharge transistor. The capacitive coupling charge current of the neighboring bit lines is then compensated for by the current of the precharge transistor.

The principle used in this test sequence can be incorporated into a more comprehensive march test to extend their capabilities in coupling fault detection.

3.9 Summary

High bit count and packing density of embedded SRAMs makes them yield limiters in SoCs. Large defect-sensitive SRAM arrays exhibit a growing number of unstable cells. We established that process variations, such as V_{TH} offset and mismatch, photo-lithography non-idealities causing L_{EFF} and W_{EFF} variations, can severely deteriorate the Static Noise Margin of an SRAM cell. The additional impact of subtle defects and extreme operating conditions can cause many cells in an SRAM array to have marginal stability and inadvertently flip their state. We described the various definitions of the SNM that can be encountered in the literature. Since analytical prediction of the effects of transistor parameter changes can be instrumental in SRAM cell design, we presented analytical expressions for the popular 6T SRAM cell as well as the 4T SRAM cells with a resistive load and the loadless 4T SRAM cell. The loadless cell analytical expression is based on the alpha-power law and tracks the simulated SNM values within 15%.

We presented an extensive analysis of the factors affecting SRAM cell stability such as the variations of the process parameters, resistive defects and operating voltage variations. Based on the learning of that analysis, the Data Retention Fault modeling using extra resistors in the pull-up path of SRAM cells has been supplemented by a Stability Fault (SF) Model. The new SF model mimics the impact of various factors that deteriorate the stability of an SRAM cell.

Based on the voltage transfer characteristics of the SRAM cell, we developed a detection concept for SRAM cells with the stability faults. It provides a better insight into the principles behind the various DFT techniques for SRAM cell stability detection including the rules for the correct setting of the pass/fail stress.

Detection of a weak cell is possible if such a cell changes state as a result of a disturbance. Repetitive read/write during a march test or resistive and/or capacitive coupling are examples of the factors that may expose a weak cell. A subsequent read operation can then detect a flipped cell. Detection capabilities of the March 11N and hammer repetitive test are insufficient for reliable detection of stability and dynamic faults, which can have the same root cause as the stability faults. More reliable test methods have to be applied to ensure a high quality stability test. We also established the conditions for the successful detection of resistive coupling faults in the neighboring SRAM columns. However, detection capability is limited to bridging defects with low resistance values.

Chapter 4
Traditional SRAM Fault Models and Test Practices

4.1 Introduction

With the advent of VLSI circuits, exhaustive functional testing has become unfeasible and has led to the appearance of structural tests aimed at detecting possible faulty conditions [16]. Such conditions have to be modeled by fault models. A fault model is a systematic and precise representation of physical faults in a form suitable for simulation and test generation [65]. Being an analytical representation of physical defects, a fault represents the functional effect of the physical imperfections on the device functionality. Note that the same functional flaw can be caused by various physical defects. Therefore, multiple physical defects can be mapped to a single fault model.

In logic- or transistor-level fault modeling all faults are assumed to be equally probable. However, in reality fault probability is a function of the probability of the defect that causes such a fault. Thus, layout-level fault modeling is essential for efficient and effective test generation that targets the defects which are more likely to appear in the layout of the circuit under test in a given technology. The defects are modeled as local disturbances in the layout of SRAM array and translated to defects in the corresponding transistor schematic. The electrical behavior of each defect is analyzed and classified, resulting in a fault model at SRAM cell level [66]. The defect modeling at layout level and extraction to SRAM cell level can be done using the Inductive Fault Analysis (IFA) [58, 67]. IFA is used in this book to obtain the SNM sensitivity to the defect resistance presented in in Section 3.5.2.

Generally, faults can be divided into time-invariant (permanent) and time-varying categories. A time-invariant fault, as is easy to guess from the term, is a fault which manifests a faulty behavior regardless of any operating conditions. If a fault manifests itself only under certain conditions or at a certain time, such a fault is often called time-varying, non-permanent, transient, parametric or reliability fault. Some of the conditions that can cause time-varying faults in SRAMs are high-energy particles, ESD events, capacitive coupling, power supply ripple and fluctuations, multiple back-to-back read or write accesses.

4.2 Traditional Fault Models

This section will touch upon the more common traditional faults found in SRAMs. Applying the reduced functional model, SRAM faults can be classified as [60]:

- Faults involving one cell:
  - Stuck-At Fault (SAF)
  - Stuck-Open Fault (SOF)
  - Transition Fault (TF)
- Faults involving two cells:
  - Coupling Fault (CF)
- Faults involving *n* cells:
  - *n*-coupling fault
  - Bridging Fault (BF)
  - State Coupling (SCF)
  - Neighborhood Pattern Sensitive Fault (NPSF)
- Stability faults
  - Data Retention Fault (DRF)
  - Stability Fault (SF)

The fault models listed above can be defined as follows:

Stuck-At Fault: An SRAM cell is designed so that it can be written a "0" or a "1" and these data values can be read back without destroying them. If for any reason a cell is stuck at a "0" or a "1" and cannot be overwritten with the complementary data value, it is said to manifest a SA0 and SA1 faults, respectively. SA fault is one of the most common faults in SRAMs and can be caused by various defects such as a short in the interconnects, transistor defects, etc.

Stuck-Open Fault: In SRAM context, a SOF means that the cell cannot be accessed due to e.g. an open word line or an open bit line.

Transition Fault: We should be able to overwrite a "0" stored in a cell with a "1" and vice versa. If, for instance, a cell fails to perform a transition from "0" to "1", but not vice versa, it manifests a TF. The state stored in the cell upon the power-up is generally a random value. Therefore, a TF may appear as a SAF if the cell assumes the state from which it cannot make a transition to a complementary state. However, unlike the SAF, a cell with a TF can be brought back into its previous state by other faults (e.g., a CF). TF can be caused by an absent access transistor.

Coupling Fault: A two-cell CF manifests itself if the contents of the victim cell is changed by a transition in the aggressor cell. A general case of a CF will involve *n* cells in addition to the two cells in the two-cell CF. In the case of the *n*-cell CF, the other *n* cells are assumed to have some fixed value as a condition of an *n*-cell CF [60]. CFs can be divided into inversion coupling faults (CFin) and idempotent coupling faults (CFid), symmetric and asymmetric faults and one-way and two-way faults. A victim cell with a CFin will flip to the other state every time the aggressor cell has a state transition. Whereas a victim cell with a CFid will flip in the same situation to a certain value (0 or 1) only. A symmetric CF causes the victim cell to undergo both high-to-low and low-to-high transition following the aggressor cell. An asymmetric CF, however, can make the victim cell to undergo either high-to-low or low-to-high transition only. A one-way CF can only be sensitized by either high-to-low or low-to-high transition, whereas a two-way CF can be sensitized by both the high-to-low and the low-to-high transitions of the aggressor cell.

Bridging Fault: A bridging fault is caused by the physical presence of excess material that "bridges" two or more lines and effectively shorts them together, a BF is often bidirectional. A victim cell is affected not by a transition of the aggressor cell, but rather by the state of the line that drives the victim cell through a conductive bridge.

State Coupling Fault: A SCF in a victim cell is caused by a *state* of the aggressor cell rather than by the transition of the aggressor cell.

Neighborhood Pattern Sensitive Fault: If the contents of a cell under test or its ability to flip states is dependent on its neighborhood, such a cell manifests a NPSF. An active or dynamic NPSF causes the cell under test to change its contents due to the change of the pattern in its neighborhood. A passive NPSF does not allow the cell under test to change its state due to a certain neighborhood pattern. And finally, a static NPSF will force the cell under test into a certain state due to a certain neighborhood pattern.

Data Retention Fault: Data Retention Fault (DRF) manifests itself if an SRAM cell fails to retain its data indefinitely while in standby (retention) mode. A cell with a DRF may pass a functional test if read-accessed again shortly. However, if the same cell is not accessed in some time, the "1" value can degrade up to the point where the defective cell flips. As the process technology continues to scale down, the DRF is becoming a greater concern. Every technology generation packs more SRAM bits in the same area. Combined with the larger transistor parameter distributions, minimal feature size transistors and the increased cell gate and subthreshold leakage, the probability of the DRF tends to increase.

Since the circuit operation speed in the scaled-down technologies tends to be limited by the propagation delay of their interconnects [55], Cu interconnects in combination with low-k dielectrics and a low-k barrier dielectric cap layer has been introduced. Most of the semiconductor foundries have switched from the traditional aluminum metallization with tungsten via plugs to copper interconnects. The lower electrical resistance of copper leads improves power distribution and device performance throughout the chip. Copper also improves the electromigration resistance, a major concern in an IC's long term reliability, by as much as 50 times.

The resistance of copper interconnects is less than two-thirds that of aluminum-tungsten interconnects. This results in RC delay reductions of 15–30% or more. Via resistance runs as low as 20% of that of tungsten plugs. These benefits become even more pronounced in 0.13 µm technologies and smaller [55]. The dual-damascene process is necessary to eliminate copper etch and for dielectric gap fill, which becomes very challenging as dimensions continue to shrink. In the dual damascene process, the trench and via patterns are defined by etching through the dielectric material in two separate lithography and etch steps. Metal filling of the trenches and vias is accomplished in a single copper electroplating step, which is followed by the Chemical and Mechanical Polishing (CMP) to remove excess overfill copper and obtain the desired metal pattern and a planar surface.

However, electromigration and stressmigration effects may limit the further scaling of copper interconnects. Increasing frequency of unreliable contacts and vias have been reported due to stress-induced voiding under the vias in CMOS 0.13 μm technology node and beyond [70] An example of a failed interconnect via is presented in Figure 4.1(b) [68].

A typical CMOS process may involve billions of contacts from metal layers to diffusion areas and vias between metal layers. Providing reliable contacts and vias is a growing challenge [68, 69], especially in high-density SRAM arrays. Poorly formed overly resistive contacts and vias can cause delay faults if located in the timing or signal propagation paths. However, our research showed that if resistive contacts are located in the load PMOS transistors of an SRAM cell, such defects can cause a DRF or a stability fault depending on their severity and can escape traditional tests.

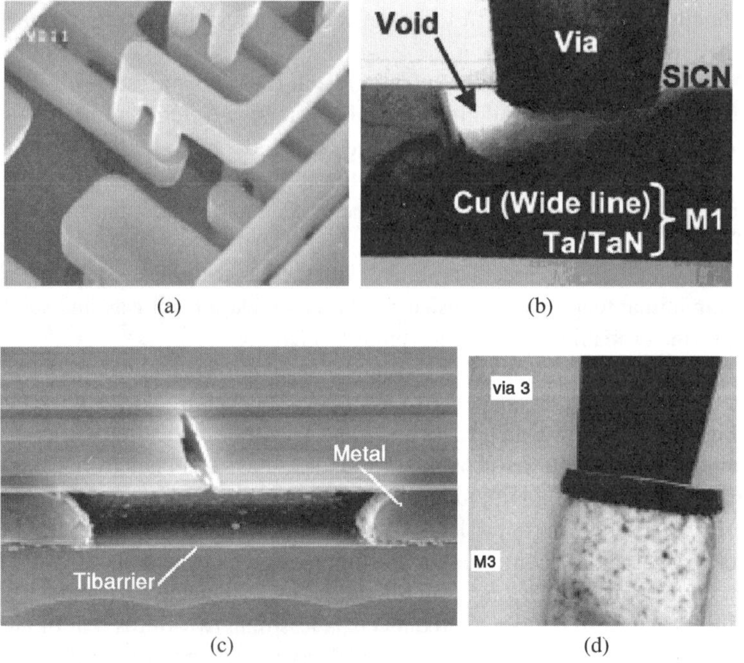

Fig. 4.1 (a) Dual damascene copper interconnects [55]; (b) Cross-sectional TEM image of a failed copper interconnect via [68]; (c) Weak open defects: detailed cross-section of a metal open line, showing the metal cavity and formation of a weak open defect due to the Ti barrier; (d) A resistive via [69]

Figure 4.1 shows dual damascene copper interconnects and illustrates weak open defects in metal and via formation. Difficult-to-detect weak opens with $R_{open} < 10\,\text{M}\Omega$ constitute a significant part of the total number of opens, as suggested by Figure 4.2. Note that weak opens are almost equally distributed across the entire range from $10\,\text{k}\Omega$ to $10\,\text{M}\Omega$ showing relatively high and flat probability of a weak open with any resistance value.

SRAM arrays are the densest form of circuitry and can occupy a significant percentage of silicon chip area. Each cell contains from ten contacts as in the cell used in this work (Figure 4.3) to 14 in the latest technologies [71]. These contacts

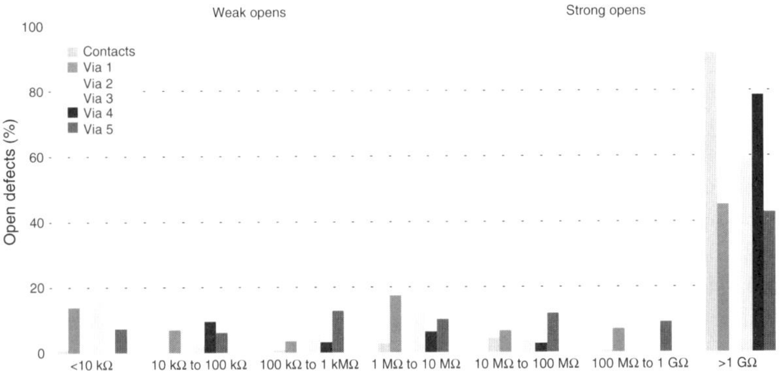

Fig. 4.2 Resistance distribution for contact and via opens [69]

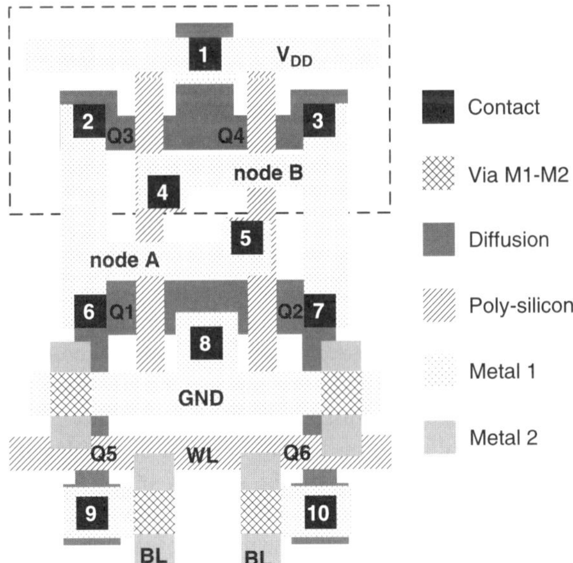

Fig. 4.3 Layout of a "tall" 6T SRAM cell, where contact resistances 1, 2 and 3 correspond to resistors R1, R2 and R3 respectively, which are shown in Figure 4.4

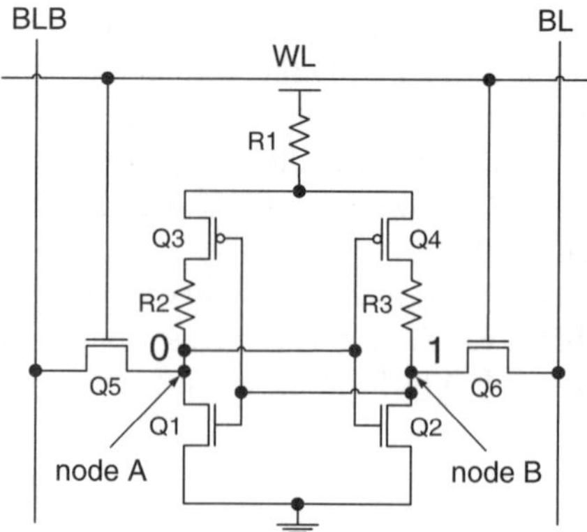

Fig. 4.4 SRAM cell schematic with resistors in place of potential weak opens that can cause stability faults as per layout in Figure 4.3

are potential locations of weak opens. A break or a weak open in contact "1" in Figure 4.3 represents a symmetric defect when both drains of load PMOS transistors $Q3$ and $Q4$ are connected to the power supply through a shared resistor $R1$ (Figure 4.4). An infinite value of $R1$ corresponds to an open in the cell's supply or to the situation when both PMOS transistors are missing. Opens in contacts "2" and "3" represent asymmetric defects in the left-hand and right-hand sides of the cell respectively and correspond to resistive connection of $Q3$ and $Q4$ sources to nodes A and B respectively (Figure 4.4). Infinite resistance in contacts "2" or "3" corresponds to an asymmetric defect [1]. As technology moves towards smaller feature sizes, the "split word line" cell layouts have been adopted by the foundries [71, 72]. They have separate drain contacts for each of the load PMOS devices increasing to four the total number of possible open contacts that can cause data retention or stability faults.

Stability Fault: The DRF that we described above are manifested by a failure of an SRAM cell to retain its value represents a severe case of cell instability and is typically attributed to a serious fabrication defect. However, if an SRAM cell just marginally passes the DRT under the typical testing conditions, the stability margin of such a cell cannot be estimated without resorting to special stability test techniques. Therefore, Stability Faults (SF) can still present themselves if the operating conditions of the SRAM array are even slightly unfavorable compared to the conditions that were set during the DRT. Let us consider a case where the stability of the cell under test is slightly compromised such that it marginally passes the DRT.

The possible SNM values of an SRAM cell can be divided in several ranges, as shown in Figure 4.5. Depending on the severity of the SNM degradation, the stability problems in SRAM cells can be classified into the Data Retention Faults and the Stability Faults with the former being a subset of the latter. For extremely low values of the SNM, the cell is likely to flip its state if not rewritten with the same data again shortly after, i.e., it fails to retain its data indefinitely demonstrating a Data Retention Fault (the leftmost range). Provided that the SNM of the cell is

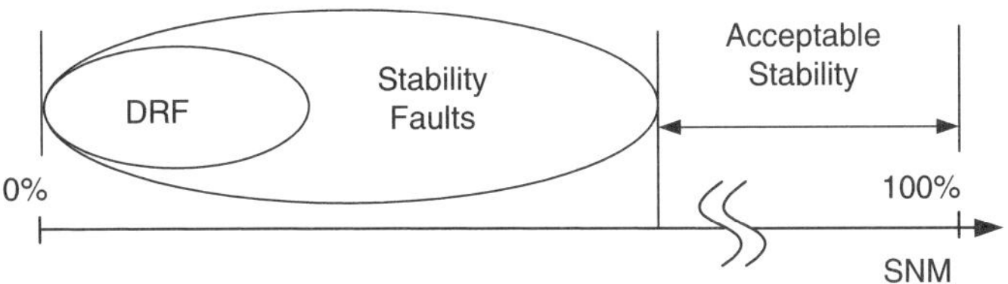

Fig. 4.5 Relationship between the SNM, Data Retention Faults (DFRs) and Stability Faults (SFs) in an SRAM cell

sufficient to handle the NMOS off-state leakage current that discharges the storage node holding a "1", under normal conditions the cell should retain its data for as long as power is supplied to the cell. However, under adverse conditions such as the reduced supply voltage, elevated temperature, increased coupling and supply noise, etc., i.e. the conditions contributing to further SNM degradation, this cell may become so unstable as to flip its state exhibiting a Stability Fault. The Stability Fault range includes the DRF range and spans on the SNM scale up to the Acceptable Stability range. Cells in the range of Acceptable Stability have the SNM that is high enough to withstand the worst possible case scenario. In Figure 4.5 they are shown outside of the oval representing the stability faults. In this book, we refer to the cells inside and outside of the oval representing Stability Faults as the *weak* and the *good* cells respectively.

A cell weakness that makes a cell vulnerable to SFs can be caused by various factors, including resistive defects (resistive breaks and bridges), excessive process shifts, mask misalignment, transistor mismatch, etc. [6]. A stability fault, which is a possible consequence of cell weakness, may occur due to any electrical disturbance such as power supply noise, read/write cell disturbs, etc. during normal operation of the SRAM. These adverse conditions, especially when combined, can cause a weak cell to flip its state easily and corrupt its contents.

The inadequate stability margin in SRAM cells affected by undetected defects and mismatches can indicate intermittent stability and possible long term reliability issues, while such cells may successfully pass regular memory tests. For certain applications requiring extreme reliability (e.g. mission-critical applications, avionics, military, banking and enterprise servers, automotive, where SoC chips control such systems as ABS, stability control, etc.) detecting all weak defects and possibly unstable cells is crucial. Stability tests are essential to achieve high product reliability and high quality tests, and should be included in SRAM test suites.

4.3 Traditional SRAM Test Practices

An efficient and economical memory test should provide the best fault coverage in the shortest test time. Besides the fault detection, manufacturing memory tests must also include a diagnostic capability that allows identification and possibly repair defective locations by applying redundant elements. In addition, the diagnostic capability can be instrumental in the manufacturing yield ramp-up by providing feedback to designers and process engineers.

The ultimate goal of testing is screening out the defective units before shipping them to the customer. As the chip complexity continues to grow, assuring a low defect count in the shipped parts can become extremely expensive. The Defect-per-Million (DPM) level can be minimized by running multiple tests that target multiple fault models. However, given the growing bit count of modern SRAMs, the time required to run nearly exhaustive test suite can be cost prohibitive. Therefore, as a test time saving approach, a test suite is often limited to include only the tests which provide the necessary fault coverage in the minimum acceptable time. The test time directly translates into the test cost. In other words, the acceptable cost of test is determined by balancing the value of test with its cost. Figure 4.6 presents the test quality trade-offs for an arbitrary chip [73].

Typically, the cost of test increases exponentially with an improvement in DPM. For certain market segments, it may become economical to lower the cost of test by using defect tolerance techniques or by accepting the increased DPM levels. However, the role of testing is not only screening the defective units. A significant value of test is realized by reducing the time to volume and expediting the fabrication yield learning. Test cost should not be minimized independently, but in the context of achieving the lowest overall manufacturing cost over a certain period of time [73].

In the following sections we will present some of the traditional SRAM test practices.

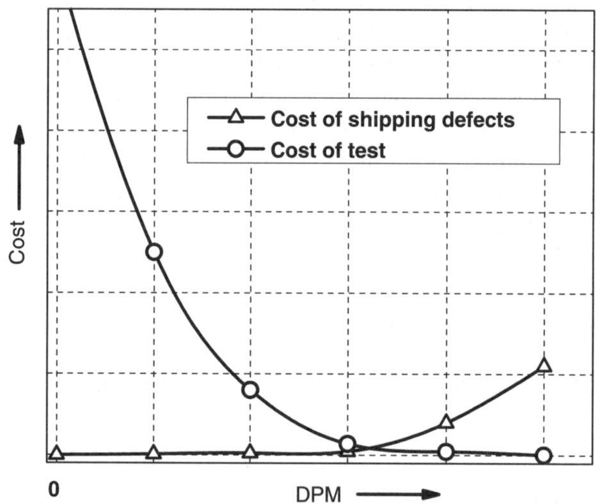

Fig. 4.6 An example of test quality tradeoffs [73]

4.3.1 March Tests

March tests are an efficient approach to ensure the correct functionality of SRAMs. March tests often target a certain fault model, such as a stuck-at, address, transition, coupling, pattern-sensitive, delay or data retention faults. The fault coverage of a different march tests varies and can range from 0% to 95–100% depending on the chosen test algorithm for every given fault. However, if the fault model poorly characterizes the real faults that may occur in a given circuit, then the developed tests can test for non-existing faults or some of the existing faults may escape the test [58]. Inductive Fault Analysis (IFA) determine which faults are more likely to occur by placing a physical defect of a given size into a particular circuit layout. Using IFA we conducted analysis of the SNM sensitivity to the resistive defects, which are likely to appear in an SRAM cell layout similar to shown in Figure 4.3 (see Section 3.5.2).

A *march test* can be defined as a sequence of *march elements*. Each *march element* contains a number of sequential *operations* which are applied to a given cell before proceeding to the next cell according to certain *address order*. The *address order* can be *increasing* (from address 0 to address $n-1$) or *decreasing* (from address $n-1$ to address 0). The increasing address order is denoted by the "$\Uparrow$" symbol and the decreasing address is denoted by the "$\Downarrow$" symbol. If the address order is not important, it is denoted by a $\Updownarrow$ symbol. Certain defects or address timing marginalities can manifest themselves only if the address increments along the row or along the column of the array. Conventionally, the row of the cell is specified by the X-address and the column of the cell is specified by the Y-address. If the direction of the address increment is along the row, the address order is commonly defined as "fast column" or "fast Y". On the other hand, if the address is incremented from row to row, it is commonly defined as "fast row" or "fast X". Among a large number of possible address sequences, the following address sequences have gained industrial acceptance: fast X, fast Y, Gray code and address complement addressing (Figure 4.7).

Fast X and fast Y with increments/decrements of 1 are effective in detecting coupling faults between topologically neighboring columns and/or rows. Fast X and fast Y addressing with increments/decrements of 2 are used to detect open faults in the address decoder paths [60]. Gray code addressing is essential for testing asynchronous memories because its address sequence contains the worst-case patterns for triggering the address-transition detection logic. Address complement is used to check for worst-case delays in the address decoding paths because all predecoder gates and the local decoder gate have to switch for each new access [74].

The fault coverage of different march tests can vary from around 40% to almost a 100% [60]. The fault coverage depends on the march algorithm and the complexity of march elements. In addition, the fault coverage is a function of multiple variables such as the design architecture, process technology, bit mapping in the array, operating conditions and many others. Therefore, the fault coverage of the same march test may vary from design to design and from technology generation to technology generation. Careful consideration should be given to defining a set of tests sufficient

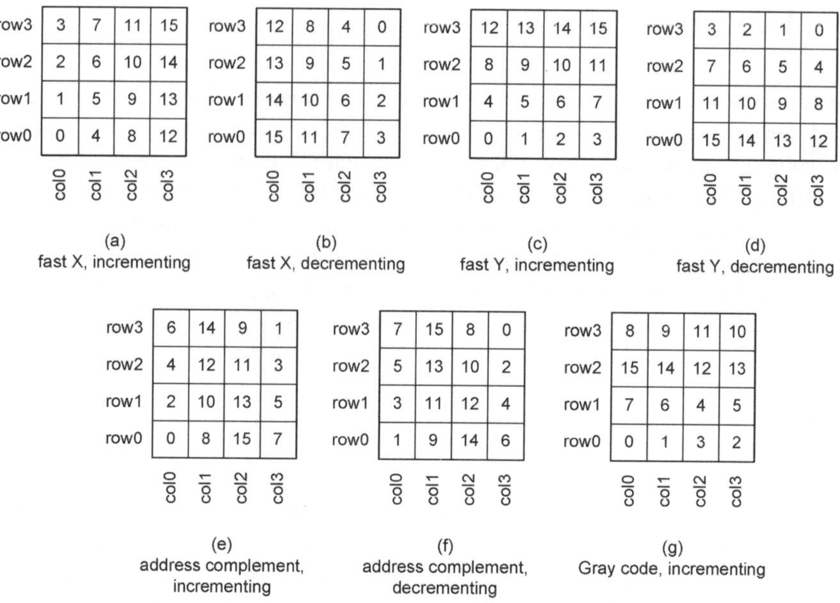

Fig. 4.7 Common addressing sequences: fast X incrementing (a); fast X decrementing (b); fast Y incrementing (c); fast Y decrementing (d); address complement incrementing (e); address complement decrementing (f); Gray code incrementing (g)

to ensure the target DPM level in each particular case. The fault coverage of the same march test applied to an SRAM array will vary for different data backgrounds and/or different address sequences [74].

March tests designed for bit-oriented SRAMs are not directly transferrable to word-oriented SRAMs. However, with a proper choice of a data background (DB), similar to those shown in Figure 4.8, a march test for bit-oriented SRAM can be used to test a word-oriented SRAM [75]. Since SRAM cells have two internal data nodes, where DB is usually referring to the left data node.

In addition to the data background, tests must account for any scrambling of the array. Scrambling means that the logical structure differs from physical or topological structure of the array. The consequence is that logically adjacent addresses may not be physically adjacent (this is called address scrambling) and that logically adjacent data bits are not physically adjacent (data scrambling). It was demonstrated experimentally that the fault coverage of a test varies by about 35% by using different addressing sequences and/or by using different data backgrounds [74]. It has become an industrial practice to apply a set of march tests several times, every time using a different pair of data backgrounds in order to compensate for the effect of scrambling [76].

There are a number of reason to introduce scrambling into SRAM arrays [74]:

- Array folding for soft-error rate and array delay minimization
- Address scrambling for address decoder area minimization
- Contact sharing between adjacent SRAM cells for array area minimization

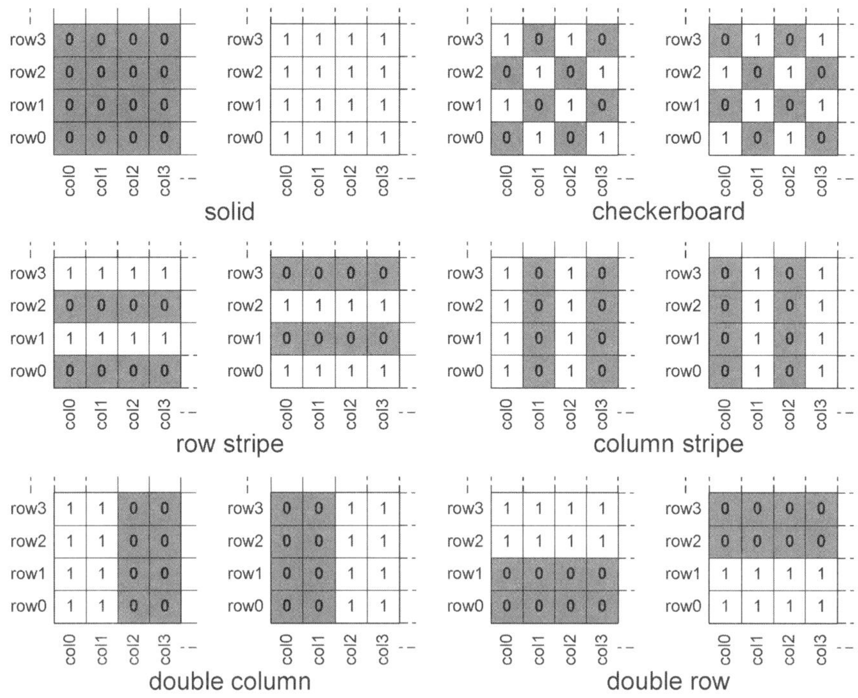

Fig. 4.8 Some common data background types

- Bit line twisting for minimization of capacitive coupling between columns
- Redundancy for yield maximization of large SRAM arrays
- I/O pin compatibility for interchangeability in commodity memories

For instance, bit line twisting reduces the capacitive coupling between the bit lines of the neighboring columns. Figure 4.9(a) shows an example of the regular column pattern without bit line twisting and the corresponding capacitances between the neighboring columns. A write operation on one of the columns that is performed by driving one of the bit lines from V_{DD} to ground can disrupt a cell in the neighboring column. If the cell in the neighboring column has a defect that causes the cell to have marginal stability, such a cell can flip and corrupt its data. Bit line twisting in the fashion shown in Figure 4.9(b) helps to reduce the coupling capacitance between the neighboring columns. In addition to achieving a greater robustness, bit line twisting improves the development of the bit line differential voltage during a read operation. By properly choosing the locations where the bit lines are twisted, it is possible to reduce the effective coupling capacitance to zero if a differential sense amplifier is used [74].

Successful sensitization of defects with a certain DB depends on the array topology. While checkerboard DB pattern is better suited for a DRAM test, the choice of the DB for SRAM is more complex. If no bit line twisting is applied (Figure 4.9(a)), a column stripe DB pattern will sensitize coupling and leakage

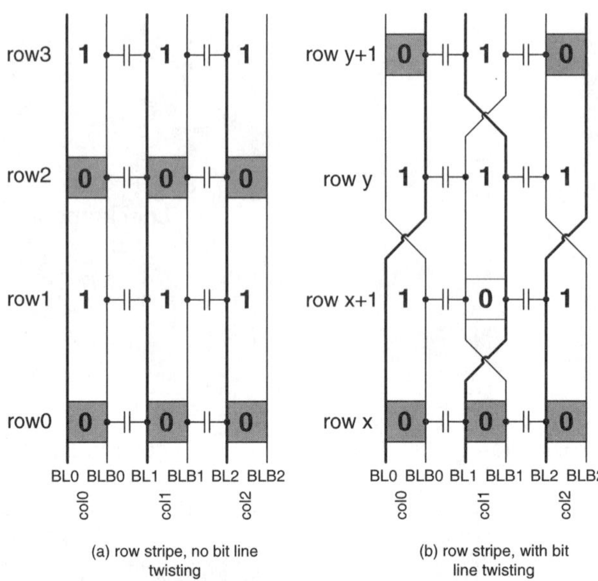

Fig. 4.9 Effect of bit line twisting on coupling capacitance and the choice of data background for defect sensitization

Table 4.1 Example SRAM march tests

Test	Sequence	N	Fault types detected			
			SAF	AF	TF	CF
MATS	⇕(w0)⇕(r0,w1)⇕(r1)	4N	+	+/-	-	-
MATS+	⇕(w0)⇑(r0,w1)⇓(r1,w0)	5N	+	+	-	-
MATS++	⇕(w0)⇑(r0,w1)⇓(r1,w0,r0)	6N	+	+	+	-
March Y	⇕(w0)⇑(r0,w1,r1) ⇓(r1,w0,r0)⇕(r0)	8N	+	+	+	+/-
March C-	⇕(w0)⇑(r0,w1)⇑(r1,w0) ⇓(r0,w1) ⇓(r1,w0)⇕(r0)	10N	+	+	+	+

effects between adjacent cells in the same row better than other types of DBs. However, if bit line twisting is introduced (Figure 4.9(b)), the DB must be modified according to the inverted internal node location.

Table 4.1 shows examples of march tests [60]. The simplest practical march test, Modified Algorithmic Test Sequence (MATS), is the basic march test that verifies only the stuck-at faults (SAFs) by writing each data background once and reading it back. By adding extra march operations/elements to the basic MATS march test, one can improve the test to cover more faults at the expense of the extra test time. The March C-algorithm is one of the most efficient march tests with respect to fault coverage/complexity. It provides the highest fault coverage detecting Stuck-At Faults (SAFs), Address Faults (AFs), Transition Faults (TFs) and unlinked idempotent Coupling Faults (CFs) [60]. More complex march tests are capable of detecting Neighborhood Pattern Sensitive Faults (NPSF). We will describe several march tests in more detail.

4.3.1.1 March C

We will describe the march elements in an example of the March C test shown in Equation 4.1. March C is effective for unlinked idempotent coupling faults [60]: unlinked inversion, idempotent 2-coupling faults and dynamic 2-coupling faults. The first march element (w0) writes a background of all "0"s into the SRAM array. The sequence of march elements of a March C test can be presented as:

$$\Updownarrow (w0) \Uparrow (r0,w1) \Uparrow (r1,w0) \Updownarrow (r0) \Downarrow (r0,w1) \Downarrow (r1,w0) \Updownarrow (r0) \tag{4.1}$$

Then, the value written in the first march element is read back during the first march operation (r0) with the expected value of "0". It is followed by a w1 operation that overwrites the same memory location with a "1". Next, the row address is incremented and the same march element (r0,w1) is repeated on the remaining rows in the same column. Address incrementing in this fashion is commonly called "fast row" or "fast X". Once the first march element has stepped through the first column, it is repeated for the next column in the same order until all columns are stepped through. Then, the third march element (r1,w0) is executed in the same row address order. The address sequencing order of the fourth (r0) march element can be either incrementing of decrementing. The addresses in the fifth (r0,w1) and the sixth (r1,w1) march elements are incremented in the opposite order. And finally, March C is concluded by reading the array values in either address order in the last march element.

4.3.1.2 March C-

Analysis shows that March C test is semi-optimal as its fourth march element ($\Updownarrow (r0)$) is redundant. March C- is an optimized version of March C. Element $\Updownarrow (r0)$ in March C- is removed without affecting the fault coverage. The sequence of march elements of a March C- test can be presented as:

$$\Updownarrow (w0) \Uparrow (r0,w1) \Uparrow (r1,w0) \Downarrow (r0,w1) \Downarrow (r1,w0) \Updownarrow (r0) \tag{4.2}$$

March C- is efficient in detecting SAFs, unlinked TFs, dynamic CFs and state coupling faults (SCFs) [60].

4.3.1.3 Hammer Test

Most of the SRAM test methods target functional fault models limited to static faults. Static faults are faults that can be sensitized by performing at most one operation. However, dynamic faults were reported based on defect injection and fault simulation of the industrial SRAMs [62]. Dynamic faults can be sensitized by more than one operation sequentially.

Equation 4.3 presents a hammer test with test length of 49N.

$$\Updownarrow(w0)\Uparrow(r0,10*w1,r1)\Downarrow(r1,10*w0,r0)$$
$$\Uparrow(r0,10*w1,r1)\Downarrow(r1,10*w0,r0) \tag{4.3}$$

The main feature of this hammer test is that the write operation is performed on the same cell ten times successively. It is denoted as $10*w1$ or $10*w0$ for a "1" and a "0", respectively. The address increment order for the hammer test was assumed "fast column" as opposed to the interlaced "fast column" and "fast row" of the 11N March test. The hammer test is classified as a Repetitive Test [61], i.e., a test that repetitively applies multiple write or read operation to a single cell. Repeating the test makes partial, hard-to-detect fault effects become full fault effects.

Dynamic faults can be caused by the same reasons as the stability faults, i.e. mostly by mismatches, defects and environmental stresses. Since only a defective cell will exhibit a dynamic fault, in the context of this book Dynamic Faults and Stability Faults can be treated as equivalent.

A successful read operation after multiple write operations requires strong pre-charge transistors, precise timing and a cell with a larger cell ratio, which translates into a larger SNM value.

Depending on the number of times each read or write operations occur in each element of a hammer test many variations of the hammer test are possible [61].

4.3.2 Parametric Testing

A serious dilemma may appear once parametric tests are applied. Many parametric tests, such as I_{DDQ} [77] that will be discussed later, raise a question of the increase in defect detection that results in a corresponding yield loss. The percentage of tested ICs that fail only certain parametric tests is a critical number when calculating the loss of revenue due to scrapping of such ICs. This percentage highly depends on the parametric test limit, the test defect coverage, and the manufacturing quality of the tested lot. These variables cause the parametric failure percentage to range from less than 1% to almost 100%. Even if a fraction of a percent of the ICs that failed only one parametric test but passed all other tests is scraped, the impact on the bottom line can be significant. Unit IC prices vary from less than a dollar to hundreds of dollars, and production volumes are often in tens of millions parts per year. The corresponding revenue loss from scrapping the parts that fail only the parametric test(s) can range from millions of dollars to over several tens of millions of dollars per year for expensive, high-volume ICs. For example, if a manufacturer produces 10 million ICs per year of $300 each and an average of 1% of these are scrapped due to the failure of only the parametric testing, then the manufacturer incurs a revenue loss of $30 million.

The yield loss challenge of parametric testing may be unsettling when ICs pass all other tests except a particular parametric test. Davidson discussed the supplier and customer trade-offs of cost versus quality [78]. He looked at IC cost as a function of warranty, repair, customer satisfaction, yield loss and test time and complexity. He concluded that for very high reliability products (e.g., automotive, satellite), shipping an IC that failed a parametric test and therefore may contain a defect is unacceptable and one cannot refer to scrapping such an IC as test yield loss. At the opposite product reliability spectrum (e.g., toys), Davidson noted that the loss of yield is a critical and overriding factor making parametric tests such as I_{DDQ} testing "untenable as a test strategy." Manufacturers whose ICs are targeting the middle-spectrum applications are facing complex quality-yield tradeoffs in deciding on the scraping threshold of the parametric tests. Possible solutions that can help to reduce the parametric yield loss include binning the ICs based on the results of parametric testing or reducing the price of the ICs that failed certain parametric tests.

However, regardless of the possible yield loss incurred by scrapping the ICs failed only parametric tests, such tests inarguably add value during the debug and characterization stages of an IC manufacturing process. Feeding the parametric test data back to design engineering and process development provides an indispensable tool that helps to improve the quality and reliability of the final product.

4.3.2.1 Burn-In Test (BI)

As was mentioned above, BI is extensively used for screening of defective chips. BI exercises the components of a chip prior to being placed into service. The BI test is often conducted at a higher temperature (up to $150°C$) and voltage (up to $1.5\,V_{DD}$) than those during the normal operation to cause accelerated aging by applying additional thermal and voltage stresses. Stressing is intended to accelerate the defect mechanisms that would have otherwise caused the early failures. However, if the die temperature is not carefully controlled, the increased voltage and heat applied during the BI may cause a positive feedback effect called thermal runaway. Thermal runaway a significant yield loss factor during burn-in [79]. It can be thought of as a destructive positive feedback condition that can occur when inadequate thermal control is combined with a silicon process technology where leakage increases exponentially with temperature [80]. Figure 4.10 shows that once thermal and voltage stresses increase the junction temperature (heating), it, in turn, causes leakages to further increase. As a consequence, it increases the static I_{DD}, which raises the power dissipation and closes the positive feedback circle that can lead to a thermal runaway.

It is well known that the failure rate over the life time of most of the semiconductor devices and electronic components in general can be illustrated by the so-called "bathtub" curve. The failure rate curve is called so thanks to its resemblance of the cross-section of a bathtub. Typically, during the initial period of an electronic device is characterized by an increased failure rate. For instance, parts that for any reason escaped the manufacturing testing but contain defects are likely to fail in the initial period of the part's life time. Once the latent defects revealed themselves, the failure rate stabilizes at the lower level. The low failure rate is maintained until

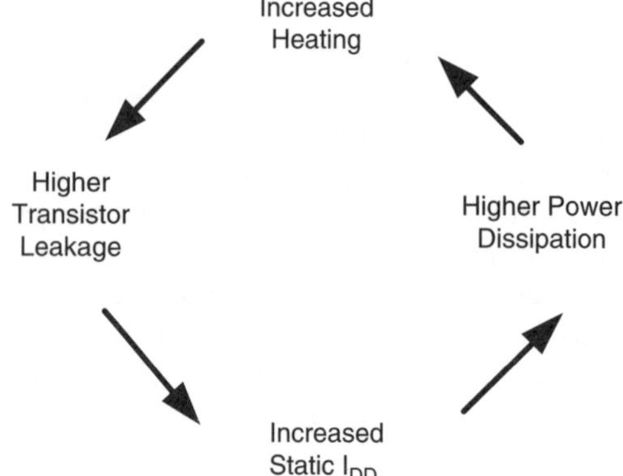

Fig. 4.10 Thermal runaway [80]

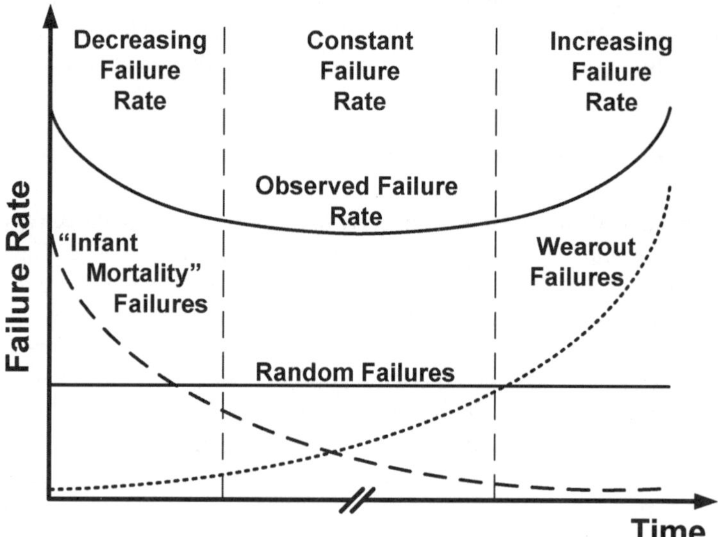

Fig. 4.11 Bathtub reliability curve

the wear-out effects do not come into play. Once the part has achieved the wear-out stage of its useful life, the failure rate starts rising again. The three stages in the product failure rate are schematically illustrated in a bathtub reliability curve is shown in Figure 4.11.

The voltage and thermal stresses applied during the BI help to accelerate the latent defect mechanisms in a device under test. Hence, if the failure rate over time of a part can resembles a bathtub, the application of BI can successfully reveal the latent defects by accelerating the aging of the device under test. The intention of the BI is to screen out the parts that would fail as a result of "infant mortality", that is, during the initial, high-failure-rate portion of the bathtub curve reliability curve. The main reasons of infant mortality are process irregularities, electromigration, surface contamination, poor moisture isolation, electrostatic discharge and random failures. If the burn-in period is made sufficiently long and extra stress is applied, the part can be accelerated through the infant mortality phase to the random failure rate phase. It can then be trusted to be mostly free of further early failures once the burn-in process is complete.

A precondition for a successful burn-in screening is a bathtub-like failure rate, i.e., the failure rate of the part should decrease after a relatively short period of overstressing it. By stressing all parts for a certain burn-in time the parts with the highest failure rate fail first and can be screened out during the Test During Burn-in (TDBI) or a subsequent functional test. The devices that survive the stress have a later position in the bathtub curve with a lower ongoing failure rate. Thus, by applying a burn-in, early in-use system failures can be avoided at the expense of a reduced yield caused by the burn-in screening process.

The constant electric field (CEF) scaling scenario requires reduction of both V_{DD} and V_{TH} to reduce to maintain a sufficient gate overdrive $(V_{DD} - V_{TH})^n$, where n varies between 1 and 2 [79]. V_{TH} scaling results in an increased leakage current. Subthreshold leakage is an inverse exponential function of V_{TH}, so that the chip leakage power increases exponentially with technology scaling. Increased leakage in turn causes the increased junction (Si lattice) temperature.

Due to increasing junction temperatures with technology scaling, the constant failure rate portion of the bathtub curve (shown in Figure 4.11) is shrinking. Figure 4.12 shows qualitatively that the wearout that is causing an increasing failure rate starts to affect device failure rate sooner as the technology is continuing to scale [80]. New technological processes, materials, reduced gate oxide thickness and other factors are hindering the estimation accuracy of optimal burn-in operating conditions (V_{DD}, temperature and time).

One has to be careful not to overuse the accelerated aging due to the BI. When the BI-accelerated aging of the device under test extends into the wear-out stage of

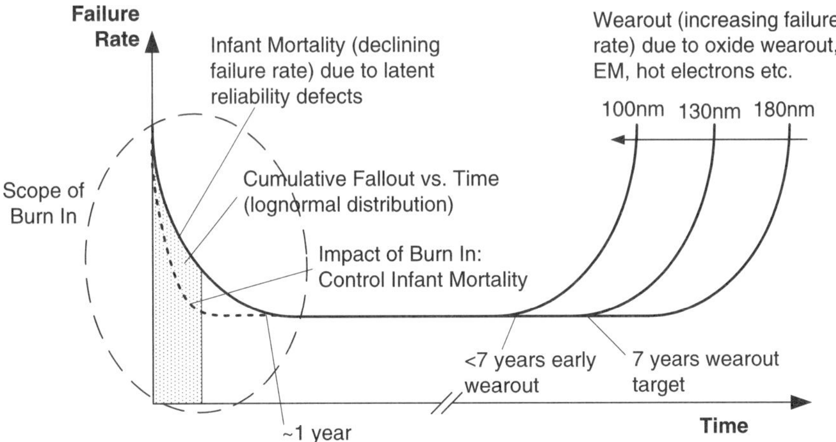

Fig. 4.12 Shrinking of bathtub reliability curve with technology scaling [80]

the bathtub-like failure rate curve, the initially positive effect of the BI will cause good parts to fail under the BI stress. The goal of the BI is to accelerate the device aging through the "infant mortality" stage only. Once the stable failure rate region is reached, any subsequent accelerated aging robs the part of its useful lifetime. However, in a mature production it is not easy to determine whether the failure rate is still decreasing or it has stabilized. To precisely determine the time distribution of device failures for a very low percentage of the production, one would have to destroy a very large number of devices, which may not be economically justified.

It is extremely important to have BI conditions optimized for a given product. If burn-in conditions are sub-optimal, the applied stress may either be not effective enough and cause higher DPM in the shipped parts or be overly strenuous and cause excessive accelerated aging which shortens the useful life of the parts. In an immature technology which is characterized by low yield, the BI is an effective defect-screening tool. Post-BI yield is often used as a good indicator to determine the efficiency of burn-in [81].

Burn-in methods include [82]:

- Whole-wafer burn-in
- Die-level burn-in
- Packaged-die burn-in

Each of the burn-in methods has advantages and disadvantages. For instance, the whole-wafer burn-in offers the most cost savings due to avoiding the packaging expenses. However, the whole-wafer burn-in will miss the defects introduced at the later production stages such as scribing and packaging. The packaged-die burn-in addresses all the manufactured steps including packaging. However, the throughput of this method is comparatively lower and the cost of screened-out parts will include the packaging cost. The die-level burn-in uses temporary carriers for each individual die and is a middle ground between the whole-wafer burn-in and the packaged-die burn-in methods.

Burn-in methods can be divided into static and dynamic. During static burn-in, a device under test is subjected to elevated temperature and supply voltages for the duration of the test. However, the DUT inputs are not toggled and only static power is dissipated. During dynamic burn-in the DUT inputs are toggled by either the burn in tester or by embedded built-in self test circuitry. Therefore, both static and dynamic power is dissipated by the DUT and voltage stress is applied to multiple components. However, due to the lower frequency (tens of MHz) provided by the burn-in tester, the dynamic power component during burn-in is significantly less than that during the normal high-frequency operation [80].

While burn-in is an effective reliability screening technique, it adds significantly to the manufacturing costs. The test time and the burn-in yield loss are the major factors. Burn-in yield is the fraction of the total number of stressed devices that meet nominal functional specifications after the burn-in. The yield loss after burn-in is introduced by the growth of the defects that is caused by the elevated temperature and voltage. Kim et al. [81] proposed a post burn-in yield loss model:

$$Y_{loss} = Y - Y_l = Y(1 - Y^{v/(1-v)}) \tag{4.4}$$

where v is the burn-in damage that depends on the temperature and voltage during burn-in. The yield after burn-in, Y_l, will be different from the yield before burn-in, Y, due to different yield critical area and average defect density after burn-in [81]. Vassighi et al. demonstrated that the post burn-in yield loss Y_{loss} is an exponential function of burn-in temperature [83]. Hence, the temperature and voltage stress must be carefully optimized and tailored for every chip exposed to burn-in conditions [82].

4.3.2.2 I_{DDQ} Test

The I_{DDQ} testing is based on the observation of the quiescent current consumption of digital CMOS ICs. Normally, a fully static CMOS gate consumes significant current only during switching. The CMOS quiescent current, however, is significantly lower than the switching current. The presence of most physical defects can raise the quiescent current by several orders of magnitude. Therefore, an increased measured quiescent current can be considered as an indication of a defect in a CMOS circuit under test.

Soden [84] (Sandia National Labs) reported I_{DDQ} qualification results on a 256 KB asynchronous SRAM in 0.5 μm technology. An asynchronous SRAM uses address transition detectors to control internal logic without an external clock. The self-timed logic of an asynchronous SRAM restricts the ability to perform I_{DDQ} test due to contention between the write driver and the PMOS pull-ups on the bit lines. Therefore, I_{DDQ} testing was performed in the disabled state of the SRAM (chip deselected). The self-timing logic prevented testing for defects, such as gate oxide shorts, in SRAM cell transistors. The contribution of a defective memory cell to I_{DDQ} current in the deselected state could be different from that while the cell is being accessed. For instance, if an access NMOS transistor has a gate-channel oxide short, while the SRAM cell is deselected, the short is not biased and therefore would not contribute to the increased I_{DDQ}. Another reported defect that is missed if an SRAM cell is deselected is a silicide bridging short across the latch transistors that can cause the cell to flip its state. Presumably, this bridge could be detected by a functional test provided that its resistance is near the critical resistance [85]; it might cause a speed- or a temperature-dependent behavior, making it hard to detect by traditional functional tests.

An SRAM I_{DDQ} test requires special test conditions [84]. SRAM array is put to a defined logic state by writing physical checkerboard patterns. Then the chip is deselected and the I_{DDQ} test is performed. These steps are repeated for a complementary physical checkerboard pattern. To ensure a better defect coverage, Sandia Lab's I_{DDQ} test for synchronous SRAMs were typically based on tests that write and read both logic "0" and logic "1" to/from every SRAM cell independently.

4.3.2.3 Limitations of I_{DDQ} Testing in Scaled Technologies

Constant-electric-field (CEF) scaling, which was adopted around the 0.8 μm technology node and beyond requires that the power supply voltage scales down as transistors dimensions are shrinking. To maintain the circuit performance, transistor threshold voltages are also subjected to scaling. These continual scaling requirements pose several technology, circuit design, and testing challenges. Controlling process variation and leakage has become critical in designing and testing ICs.

Generalized constant electric field (CEF) and constant voltage (CV) scaling relationships are summarized in the Table 4.2 [86].

In the Constant Voltage (CV) scaling the electric-field pattern of the scaled transistor is preserved, but the intensity of electric field is increased by a *scaling factor* ε. Hence, the voltage is scaled by ε/S and can be decreased much slower along with the threshold voltage. However, this approach is limited by reliability effects such as hot-carrier degradation and gate oxide failure. Another very significant limit to ε is the power dissipation. In order to maintain the same maximum speed, the power dissipation increases by ε^2 as given by CV^2f.

Scaling the threshold voltages of the devices down along with the supply voltage exponentially increases the standby leakage current (I_{off}) of the circuit [87]. The growing leakage currents and the power, which is wasted for dissipation of the generated heat is limiting the practical power supply scaling [88]. Process variations together with low-V_{TH} devices can significantly increase the absolute leakage magnitude. The die-to-die and intra-die parameter variations are also worsening with technology scaling. These variations affect the maximum clock frequency and leakage power distributions. The variation effects are more pronounced at low supply voltages (V_{DD}).

Technology scaling also affects various aspects of SRAM testing. Specifically, elevated transistor leakage and excessive parameter variations in scaled process technologies threaten the feasibility of leakage-based tests such as I_{DDQ}. I_{DDQ} testing in its traditional form has been successfully used for defect screening of FETs with the channel length of 0.25 μm and above [52].

Researchers have proposed several methods for adapting I_{DDQ} testing to scaled technologies. Motivated primarily by the increasing adverse device leakage trends, researchers have recently reported several methods for sustaining the effectiveness of I_{DDQ} testing for scaled technologies among them are reverse body bias (RBB), current signatures, ΔI_{DDQ} testing, and transient current testing. Applying a reverse body bias creates a low-leakage I_{DDQ} test mode [89, 90]. Gattiker and Maly

Table 4.2 Generalized scaling relationships

Physical parameters	CEF scaling	CV scaling
Linear dimensions	$1/S$	$1/S$
Electric field intensity	1	ε
Voltage	$1/S$	ε/S
Channel doping	S	εS

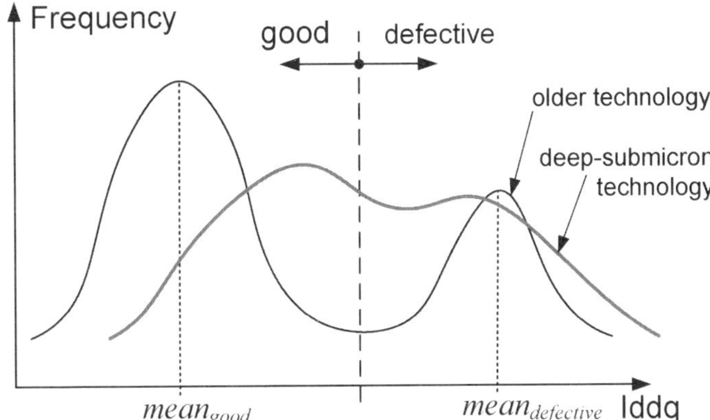

Fig. 4.13 A comparative I_{DDQ} general distribution trends seen in CMOS circuits in older technologies and in modern deep-submicron technologies

suggested sorting I_{DDQ} test vectors in ascending order; an abrupt discontinuity in the current level indicates a defect [91]. Maxwell and colleagues demonstrated the effectiveness of current signatures with silicon data [92]. Thibeault and Miller each proposed a ΔI_{DDQ} test technique for uncovering defects [93, 94]. Conceptually, the I_{DDQ} technique is similar to the current signatures technique, where a sudden elevation in current level indicates a defect. Other proposals involve transient current test techniques [65, 95]. Keshavarzi et al. proposed a correlative multi-parameter I_{DDQ} test [96]. Since device leakage and its switching speed are functions of the threshold voltage (V_{TH}) and transistor channel length (L_{eff}), a strong correlation can be established between the I_{DDQ} and maximum operational speed f_{max}. A linear dependency of the leakage current on the f_{max} was observed across the range of natural variation in transistor and circuit parameters. The multi-parameter approach does not rely on the absolute value of the current. Added parameters of temperature and body bias further improve the defect sensitivity of such a test technique. Leakage averaging and variance reduction techniques, such as nearest-neighbor residual, may mitigate parameter variation limitations. Using the adaptive body bias technique was proposed to enhance manufacturing yield by compensating for die-to-die and intra-die parameter variations and their effect on a circuits leakage and frequency. However, the body biasing techniques are becoming less effective with every new technology generation.

As the technology continues scaling in the nanometer region, the increasing leakage and process variations further reduce the usability and accuracy of I_{DDQ} testing.

4.3.3 Design For Test Techniques

In general, DFT techniques seek to improve the testability by including additional circuitry. While the additional DFT circuitry can increase design cost, it is often offset by the improvement of the quality level. For certain kind of testing DFT can be a preferred or even exclusive test solution. For instance, thc testing capabilities offered by SRAM cell stability test techniques described in Section 5.4 which can be classified as the DFT techniques can only be achieved by the addition of additional test circuitry. DFT circuitry can be controlled either directly by the external tester (ATE) or through a Built-In Self Test circuitry described below.

4.3.3.1 Built-In Self Test

Built-In Self Test (BIST) is a special DFT technique that facilitates internal test pattern generation and output response compaction [15]. A basic BIST architecture is shown in Figure 4.14. Two essential parts of a BIST are the Test Pattern Generator (TPG) and the Output Response Analyzer (ORA). The TPG generates a sequence of patterns for testing of the CUT, whereas the ORA compacts the circuit responses into some type of pass/fail indication. Test Controller provides control signal to all blocks of a BIST. The input isolation circuitry switches the input of the CUT from the normal system inputs to the TPG outputs.

Applying BIST strategies is often the only economical way to ensure at-speed testing and to overcome the limited observability and controllability of embedded SRAMs.

Running at-speed test helps to identify delay faults in manufacturing and burn-in tests that otherwise might not be detected in a slower speed test provided by ATE. Moreover, BIST-equipped SoCs are less demanding to ATE's test vector memory and number of serviced pins. Thanks to the relaxed driving speed and vector memory requirements, PBIST-equipped SoCs may be tested at less expensive testers. The extra design time incurred by including BIST is often offset by the savings in test development time and can expedite time-to-market in some cases. Disadvantages of BIST circuitry include the area overhead, design effort and performance penalties of the circuit isolation MUXs in high-performance designs. However, in most cases the benefits of BIST overweigh its disadvantages.

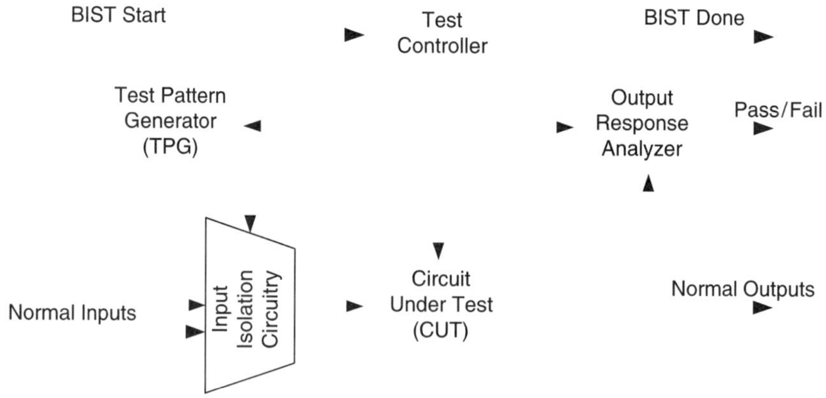

Fig. 4.14 Basic BIST architecture [15]

4.4 Summary

Exhaustive functional testing is unfeasible in modern VLSI technologies. Therefore, functional testing has been succeeded by the structural testing that relies on the fault models representing physical defects. The probability of each kind of a defect to appear in a given circuit and cause a certain fault is strongly layout- and technology-dependent. We presented a number of SRAMs-specific fault models as well as more common fault models regarded in an SRAM context. A detailed overview of the traditional SRAM test techniques including march tests and parametric tests is presented. The common address sequences and data background used in march tests as well as fault coverage of various march tests is discussed followed by the trade-offs between the yield loss and the added quality level introduced by the parametric tests. The chapter concludes with a short overview of the benefits Design-for-Test techniques on an example of the BIST.

Chapter 5
Techniques for Detection of SRAM Cells with Stability Faults

5.1 Introduction

Detection of stability faults in SRAM cells by the means of the functional tests can be a time consuming and expensive effort. Except for the most severe cases, SRAM cells with reduced stability may not be readily detected by the traditional memory tests. Detection faults such as the Data Retention Fault by the traditional memory tests requires extra pause periods of the order of hundreds of milliseconds for each data background used in the test sequence and/or increased temperature. The increased test time, partial test coverage and extensive silicon characterization to determine the worst-case test conditions add to the list of the drawbacks associated with the use of the traditional tests for cell stability test in SRAMs. Both the extra test time and/or the high temperature requirements add to the test cost and may become prohibitive. The SRAM I_{DDQ} techniques, which were popular in earlier technology generations, suffer from the reducing of the diagnostic resolution in the new technology generations and fail to detect any type of data retention faults or stability faults in on an individual memory cell basis with any degree of accuracy [1].

5.2 Classification of SRAM Cell Stability Test Techniques

Practical SRAM cell stability test techniques employed in the industry can be classified into passive and active based on whether or not special DFT circuitry is required to apply the test stress to an SRAM Cell Under Test (CUT). In turn, the active or DFT techniques can be divided into techniques with a single test stress and techniques with programmable (multiple) test stresses. Classification of SRAM cell stability test techniques is presented in Figure 5.1.

As a rule, the passive test techniques do not require special circuitry to create the test stress and might not be able to assert elevated levels of stress on the cell.

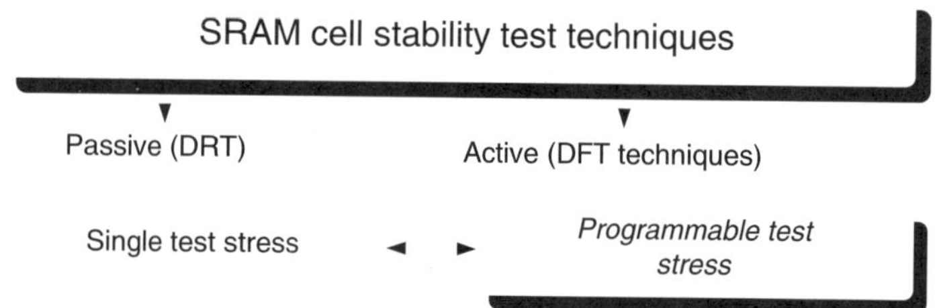

Fig. 5.1 Classification of SRAM cell stability test techniques

The effectiveness of passive tests relies on conditions under which a healthy cell is designed to be functional, including the data retention and cell stability functions. While such conditions are less likely to be met by a traditional algorithmic tests, they can occur during the normal operation of an SRAM array in a final product.

In contrast, the active or DFT test techniques typically require additional special circuitry which enables applying elevated levels of stress to the cell under test. Such elevated stress levels might occur during normal operation of a product, but are unlikely to occur during the traditional algorithmic tests. Active cell stability test techniques with a single test stress setting apply a fixed level of stress to the cell. Recently, the single test stress techniques were complemented by programmable test stress techniques. The programmable test stress techniques offer extended flexibility in choosing (programming) the applied test stress to the cell. The benefits, drawbacks as well as examples of various kinds of SRAM cell stability test techniques will be discussed below.

5.3 Passive SRAM Cell Stability Test Techniques

The data retention test and low-voltage test is an example of passive SRAM cell stability test techniques. These techniques do not require extra test circuitry. Instead, they rely on extended pause times in the data retention test and on the tendency of defective SRAM cells to lose stability at the reduced power supply voltage faster than a healthy cell would [97].

5.3.1 Data Retention Test

Section 4.2 presented the Data Retention Faults (DRF) and the physical reasons for their formation. In this section, we present the design and test aspects of the Data Retention Test aimed at the detection of the DRFs.

Unlike the dynamic RAM cells, static RAM cells are designed such that the data is retained in a cell as long as the cell is powered up without a need for refreshing the data at a fixed time interval. However, the data retention is guaranteed only in a defect-free SRAM cell and provided that the power supply voltage is above a certain value. The traditional method for detecting such data retention defects provides a tester delay interval long enough to allow the node carrying a "1" in a defective SRAM cell to discharge. That is, it increases the delay interval between the write of the test data pattern to the SRAM and the subsequent read-verify of the SRAM. Unfortunately, such tester delays significantly increase the time required for testing each integrated circuit. Consequently, more integrated circuit testers are necessary to obtain a certain throughput of integrated circuits when a data retention test with such a delay interval is used. The increased need for the expensive integrated circuit specialized testers results in an increase in manufacturing costs.

Growing V_{TH} fluctuations have significant detrimental impact on the cell stability even in a defect-free SRAM cell due to the increasing effect of random distribution of dopant atoms in deep-submicron transistors of SRAM cells [41] and gate-oxide soft breakdown [98] causing excessive gate-diffusion leakage. If in addition to the process fluctuations, a cell contains a defect that reduces the pull-up current of the PMOS transistor, the likelihood of a retention fault increases dramatically. In less severe cases an SRAM cell can develop an intermittent stability fault. The pull-up PMOS transistor, as well as the access NMOS transistor in a 6T SRAM cell is typically sized with minimum width to enable higher array density [99]. Therefore, random dopant fluctuations affect these transistors the most endangering the reliability of data retention.

Figure 5.2(a) shows a 6T SRAM cell in retention (quiescent) mode when $V_{BL} = V_{BLB} = V_{DD}$, $V_{WL} = 0$, where $Q1$ and $Q2$ are driver transistors, $Q3$ and $Q4$ are load transistors, and $Q5$ and $Q6$ are access transistors, respectively.

Poorly formed contacts in the cell shown as $R1$, $R2$ and $R3$ in Figure 5.2(b) can cause SFs in an SRAM cell. A break or a weak open represented by $R1$ creates a symmetric defect when both drains of load PMOS transistors $Q3$ and $Q4$ have a resistive connection to the power supply. Infinite value of $R1$ corresponds to an open in the cell's supply or to the situation when both PMOS transistors are missing. Opens represented by $R2$ and $R3$ create an asymmetric defect in the left-hand and right-hand sides of the cell, respectively, and correspond to the resistive connection in the pull-up path with either transistor $Q3$ or $Q4$.

A *Data Retention Fault* (DRF) is defined as the failure of an SRAM cell to retain the written data for as long as the power is supplied. Figure 5.2(b) helps to explain the conditions causing a DRF. If the off-state current I_{off_Q2} of transistor Q2 in Figure 5.2(b) is such that

$$I_{off\_Q2} > I_{pull-up} + I_{off\_Q6},$$

where $I_{pull-up}$ is the current in the pull-up path of a cell and I_{off_Q6} is the off-state current of Q6, then after a delay proportional to

$$C_B V_B / (I_{off\_Q2} - (I_{pull-up} + I_{off\_Q6}))$$

the capacitance of node B (C_B) will discharge sufficiently for the cell to flip states. Reading the cell data after a delay on the order of 100 ms and comparing it with the

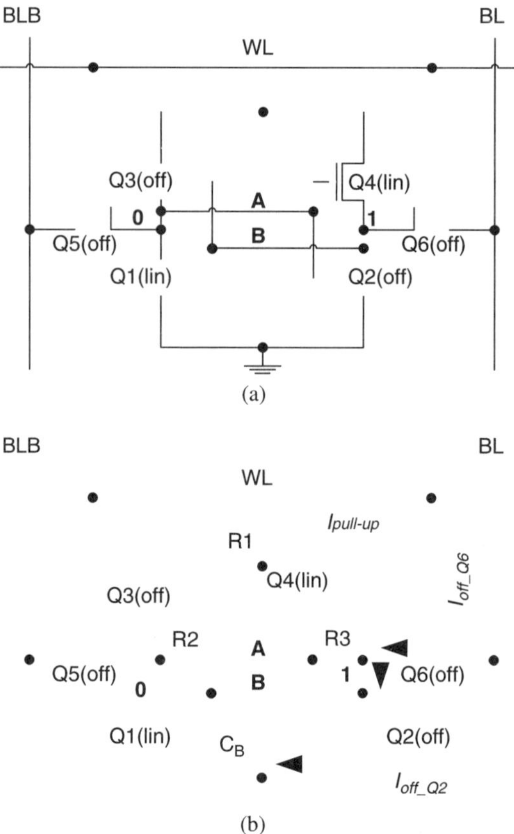

Fig. 5.2 (a) Defect-free SRAM cell 6T SRAM cell in retention (quiescent) mode when $V_{BL} = V_{BLB} = V_{DD}$, $V_{WL} = 0$; (b) SRAM cell with a symmetric (R1) and asymmetric (R2, R3) defects in data retention mode

previously written data can detect resistive defects R1–R3 in the range of several GΩ. This algorithm is employed by the traditional passive Data Retention Test (DRT or Delay or Pause Test).

Conditions and the defect resistance range of a DRF detection by means of the Pause (DRT) test are illustrated in Figure 5.3. The DRT test reads the SRAM array after a pause on the order of 100 ms to determine whether the content of any cell has changed from its previously written state [1]. This pause time should be multiplied by two to account for the pause time that is spent during the DRT of the CUT with a complementary data. This prolonged pause during which the part is occupying an expensive memory tester significantly contributes to the test cost and thus to the overall cost of the part.

The DRF in the cell is modeled by an asymmetric defect resistance $R3$ (Figure 4.4). If $R3 = 50\,\text{M}\Omega$, then such a highly resistive defect is not detected even at the elevated

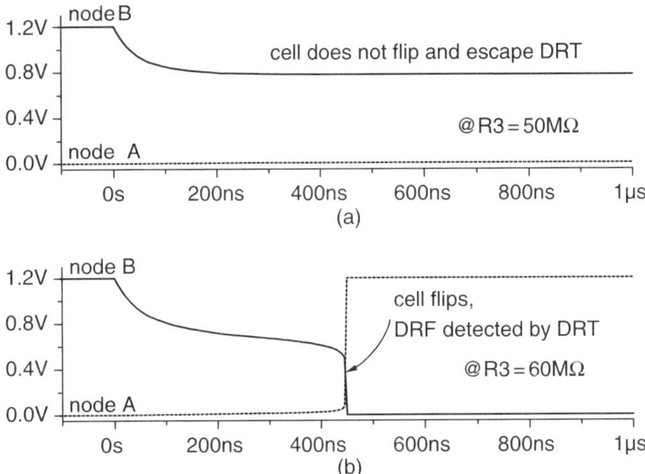

Fig. 5.3 Data Retention Fault due to the discharge of node B through the off-state current of Q2 (simulation results for CMOS 0.13 μm technology, V_{DD} = 1.2 V, T = 150°C). A resistive open R3 = 50 MΩ is insufficient to flip the cell and thus is not detected by Data Retention Test (DRT) (a), whereas R3 = 60 MΩ causes a DRF and is detected by DRT (b)

temperature of 150°C and a pause of more than 100 ms. When $R3$ = 60 MΩ, the off-state current of $Q2$ is sufficient to gradually discharge the capacitance of node B and at 450 ns the cell flips, destroying the stored data. If the same test is conducted at room temperature, the lowest detected value of $R3$ will be 2.75 GΩ and the test time necessary to detect it will be over 60 times longer. Obviously, the detection range of the DRT is insufficient to reliably identify many manufacturing defects that cause poor cell stability. Elevated temperatures improve the detection range by about 45 times at the cost of the increased test time. However, resistive opens of about 50 MΩ are still considered to be strong opens [69]. Unless special tests are applied, such defects will pass the standard test and an SoC with highly unstable and unreliable SRAM cells will be shipped to the customer. More subtle defects caused by the process disturbances can also reduce the stability of the cell. The likelihood of resistive bridges grows as the critical area shrinks with scaling, and resistive break defects are likely to appear in place of poor or absent contacts, vias or silicide [1,2]. While these defects can be non-catastrophic, they can have a serious impact on the cell stability as shown in Figures 3.20 and 3.21, respectively.

A typical Data Retention Test (DRT) is implemented as a pause of an order of 100 ms between the march elements. DRT can detect a complete open in the pull-up path of an SRAM cell. In case of a symmetric defect, where the pull-up paths in both the inverters are open, the detection is not dependent on the data value stored in the cell. However, if a cell has an asymmetric defect, where only one of the inverters has an open in the pull-up path, the DRT will only detect an open in the pull-up path of the node storing a "1" [60]. This property of the DRT requires to run it for each

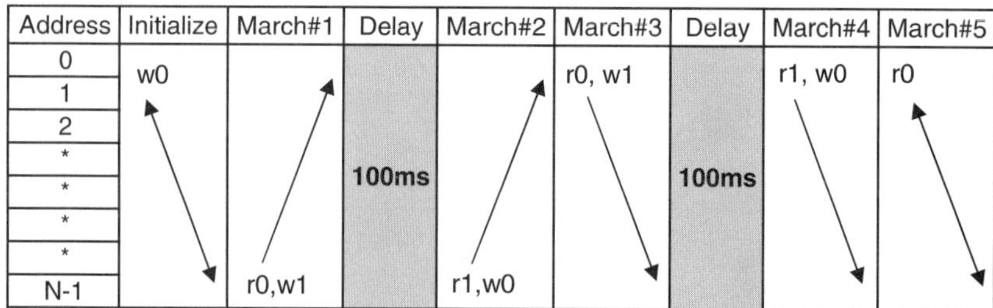

Fig. 5.4 Inserting delay elements into the March C- test to include a Data Retention Test (DRT)

of the two opposite backgrounds to cover both the asymmetric faults in the pull-up path of the cell.

Since running the DRT for each of the data backgrounds takes 100–200 ms [1], the DRT will cause each chip to spend an extra 200–400 ms on the tester. Figure 5.4 illustrates the insertion locations of the Delay elements between the march elements March#1 and March#2, and March#3 and March#4 in March C- test. In many cases, the total impact of using the DRT on the test time and hence, the test cost can be prohibitively high. Combined with the limited defect resistance coverage of the DRT and the high-temperature requirement to improve it, the disadvantages of using the DRT urge the design and test engineers to replace it with special Design for Testability (DFT) techniques. Some of the DFT techniques for SRAM cell stability detection will be presented in Section 5.4.

5.3.2 Low-Voltage Test

SRAM cell stability is one of the key constraints in our ability to continue the constant electric field scaling (CEF) of semiconductor technology. In order to maintain CEF scaling without significant modifications of the transistor structure, the scaling of the transistor dimensions has to go hand in hand with reducing the V_{DD}. However, as was shown in Section 3.5.3, SRAM cell stability is directly proportional to the V_{DD}. Monitoring the minimum voltage V_{DDmin} at which the cell is still functional is a common metric in process/design co-optimization for a given application [34]. An ideal cell should work at extremely low values of V_{DD}. However, in reality the process nonidealities and transistor parameter mismatches and offsets can either cause a cell to flip and lose its data during a read access or can render it unwritable when the supply voltage of the cell drops below a certain V_{DDmin} value. This problem is more pronounced as the array sizes grow because the statistical SNM distribution in large SRAM arrays is also becoming wider as the array size grows. The V_{DDmin} is limited by the worst cells in the array, while the probability of encountering such cells in large SRAM arrays can be significant. Therefore, defining V_{DDmin} as the V_{DD} value where SNM of a typical SRAM cell equals zero can be too optimistic for arrays with significant bit counts. Since the SNM of the worst cell in the array can statistically be significantly lower than that of a typical one, this worst cell will limit the low-voltage performance of the entire array.

V_{DDmin} data can be correlated with the parameters of the cell transistors and fed back to fine tune the fabrication process in the effort to minimize parameter variations and the manufacturing defect density and thus, the V_{DDmin} of an SRAM array.

As back-end process technology migrated to the Cu dual-damascene process in advanced technologies, metal residues and particles are no longer the key failure mechanism in SRAMs [100]. Instead, the gate oxide leakage caused by oxide thickness scaling has been reported as a significant problem [101]. For instance, the increased gate oxide leakage of the NMOS pull-down transistor of the node storing a "0" will work against the pull-up current of the PMOS transistor on the "1" storage node. Therefore, an additional strain is put on the PMOS transistor and interconnect quality and process tolerances to maintain node which stores a "1" above the switching threshold of the other inverter of the CUT under all operating conditions. Every new process generation creates more technological challenges. For instance, random doping fluctuations in nano-scaled MOS transistors [35]. They are caused by the physical dimensions of a transistor becoming commeasurable to the granularity of the transistor material lead to larger parameter variations of such transistors. This and other factors contribute to large variations in transistor's V_{TH}, I_D and leakage. Any asymmetry in the relative strength of the two inverters in an SRAM cell caused by transistor variations is a strong factor that contributes to low SNM values leading to an unstable cell.

Therefore, the bulk of the V_{DDmin} distribution across the SRAM array provides vital clues on overall SRAM cell design and process optimality and cell stability, while the tail of the V_{DDmin} distribution indicates the quality level of the manufacturing process. Thompson et al. [102] shown that the tail of the V_{DDmin} distribution is sensitive to defects that can be can occur within the bitcell during processing. Such defects may contribute to unwanted leakage paths.

Figure 5.5 shows the leakage currents in a 6T SRAM cell [103]. The magnitude of leakage currents may be high enough to affect the measured V_{DDmin} but still low enough to not impact the overall functionality of the memory. However, such defects may still impact the long-term reliability of the die containing the memory. Large transistor parameter mismatches and the gate leakage may cause the node storing a "1" to gradually discharge due to the increased gate voltage V_G of the off-state pull-down transistor (Figure 5.5) and if the discharge continues, the cell can lose its data [80]. Therefore, it is necessary to monitor V_{DDmin} and minimize the presence of outliers in the V_{DDmin} distribution in order to achieve high manufacturing quality.

An example of a defect that can impact the V_{DDmin} distribution is a bridging defect between the internal nodes of the cell as shown in Figure 3.27 in Section 3.6. If the resistance of such a bridge defect is sufficiently low, then the stability of the cell is lowered enough such that the state of the cell can be flipped during the read operation. Since the SNM is directly proportional to the voltage supply (see Section 3.5.3), the SNM deterioration due to defects would have to be compensated by an increase in the supply voltage, i.e. an increase in V_{DDmin}.

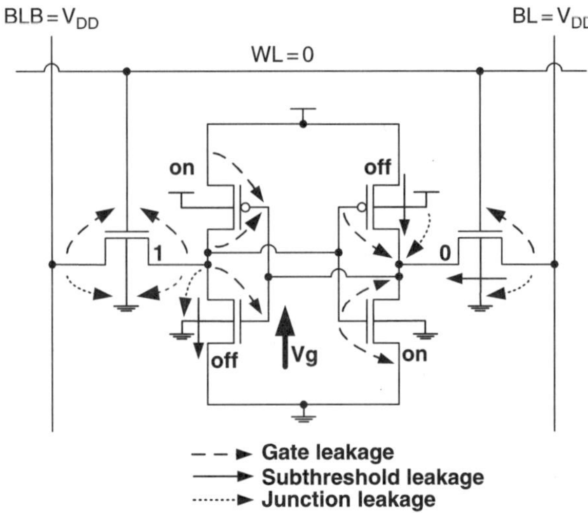

Fig. 5.5 6T SRAM cell showing different leakage components

The functionality of the memory as a whole is often dictated by one (or few) rogue bit(s). This bit fails below the measured V_{DDmin} value of the array whereas the remaining bits in the memory are functional over a significant range of voltage below V_{DDmin}. As the voltage supply is further reduced, additional bits start to fail due to the peripheral circuitry that starts failing at low voltages. Venkatraman et al. reported that physical analysis confirmed the existence of the physical defect that created the rogue bit [34].

Physical analysis is often used to systematically identify process-related defects that can negatively impact the V_{DDmin} distribution. As process technology matures, the process optimizations results in a tighter V_{DDmin} distribution with a lower mean and fewer outlier bits. Note that the minimum value of V_{DDmin} of an SRAM chip can also be limited by the V_{DDmin} of the peripheral circuitry as opposed to by the V_{DDmin} of bitcells.

Recently, an observance of erratic V_{DDmin} behavior was reported [104]. Investigation of the root cause of this erratic behavior pointed at the Random Telegraph Signal (RTS) noise that was induced by the soft breakdown of the gate oxide of the pull-down NMOS. The RTS was caused by electron trapping/detrapping in the transistor's gate. The erratic behavior was shown to be attributed to a single erratic bit moving between the tail and the bulk of the V_{DDmin} distribution, while the location of the fluctuating bit can be spatially random. The tail of the V_{DDmin} range distribution increases as the number of readouts is increased. Observation of erratic fluctuations is directly dependent on the amount of sampling. The erratic behavior was modeled by a resistor between the gate of one of the pull-down NMOS transistors in an SRAM cell and the ground (Resistor R_g in Figure 5.6). Resistor R_g represents the gate-to-source leakage of the NMOS transistor.

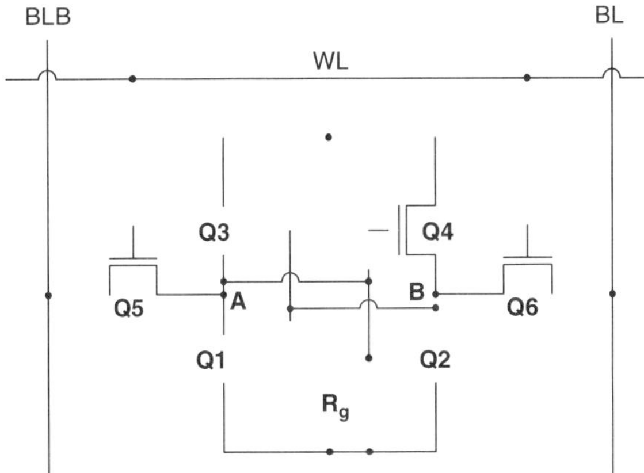

Fig. 5.6 Erratic bit fault modeled as a resistor R_g representing NMOS gate leakage [104]

The V_{DDmin} behavior is reported to be sensitive to the gate leakage of the NMOS pull-down transistor and indicated that small fluctuations in gate leakage due to the RTS noise could lead to large fluctuations of V_{DDmin}. The V_{DDmin} sensitivity to R_g increases for higher levels of the gate leakage. Erratic V_{DDmin} behavior intensifies as the cell size is reduced and thus represents another constraint on the scaling of SRAM cells and on the minimum operating voltage of the SRAM array. An improvement in the gate oxide quality can help combat the erratic bit behavior. Another measure that is reported to mitigate the erratic bit phenomenon is the optimization of SRAM cell device strength ratios.

The stress and aging effects of burn-in (BI) [82] can degrade the transistor parameters, disrupt the optimality of the cell design and cause an increase in SRAM V_{DDmin}. BI can thus serve as an additional tool that helps to screen out SRAM cells with elevated leakage, transistor mismatches and manufacturing defects.

Due to simplicity and test capabilities, low-voltage test gained industrial recognition. For instance, virtually all process development vehicles beyond 0.18 μm at Taiwan Semiconductor Manufacturing Company (TSMC) are exercising the low-voltage test mode [2]. The exact screening criteria for V_{DDmin} depends on the trade-offs between the array area, power and access speed, which are tied to product application and reliability requirements [100]. For instance, these requirements may significantly vary depending on whether the product is to be used in a high-performance or a mobile application.

5.4 Active SRAM Cell Stability Test Techniques

In an effort to extend the detection capabilities and cut test costs, passive SRAM cell stability test techniques were largely replaced by a number of structural stability test methods. The active SRAM stability test techniques offer multiple benefits over the functional tests such as the DRT. The active tests enable significant test time reduction, improved test quality, manufacturing flexibility and productivity. Unlike the passive DRT, the active stability test techniques normally do not require high temperature and therefore can be effectively used for wafer testing. The active SRAM stability test methods apply stress the cell to determine the degree of its stability. The general idea of active stability test techniques is to detect the defects that can result in stability faults, i.e., the causes of a faulty behavior rather than testing the retention time, i.e., the symptoms of a faulty behavior.

Based on the flexibility of the test stress setting, active or DFT-based SRAM cell stability test techniques can further be divided into two groups: (a) techniques with a single test stress level or pass/fail threshold and (b) techniques with variable/programmable multiple pass/fail thresholds.

A critical part of designing any circuitry for testing of SRAM cell stability is the robustness of the DFT circuit itself. It is very important to maintain the set pass/fail threshold during the stability test such that the test stress will not fail the good cells. However, the definition of a good cell is often not trivial. It is dictated by the SNM distributions observed in the utilized process technology as well as the quality/cost targets of the final product. A general drawback of the active stability test techniques is the area and the design costs associated with the additional test circuitry.

Below, we will describe a number single and programmable active test techniques for SRAM stability test in more detail.

5.4.1 Active Techniques with a Single Test Stress Level

Early active methods for SRAM cell stability test offered a single stress setting, which was determined by the best estimates of pre-silicon process conditions. In this section we will describe a number of active stability test techniques with a single, or non-programmable, test stress level.

5.4.1.1 Soft Defect Detection

One of the first test techniques for subtle defects dubbed "Soft Defect Detection" or SDD Technique was proposed by Kuo et al. in the late 1980s [3]. This pioneer stability test technique targeted the data retention capability of the 6T CMOS SRAM cell which was gaining popularity. The SDD technique performed a retention test at room temperatures. Moreover, it could also detect the subtle defects that were previously impractical or even impossible to detect using the conventional memory tests. Kuo defined *soft defects* as those that can cause unpredictable data retention failures that are process, temperature and time-dependent. In contrast, defects that can easily be detected by the functional testing, such as shorts, are defined as *hard defects*.

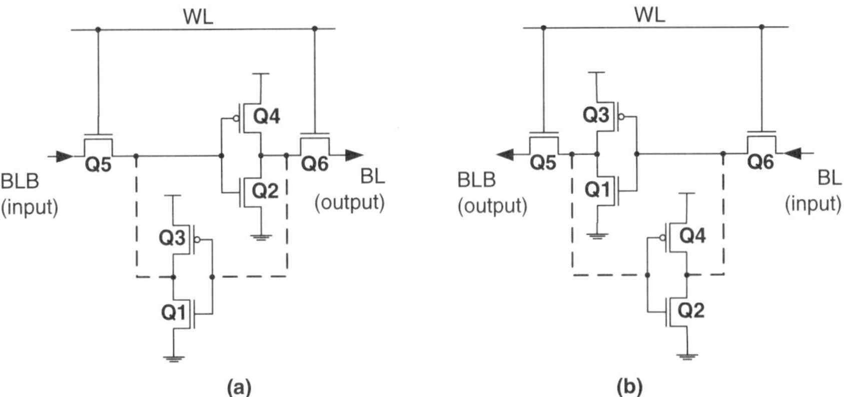

Fig. 5.7 The representation of an SRAM cell as two separate inverters for SDD open-circuit test: right inverter (a) and the left inverter (b) [3]

The SDD technique was reported to detect all possible open defects in an SRAM cell by performing a complete continuity check including the gate connections for the floating-gate transistors. This can be accomplished by two tests: the SDD test for opens in the cell and the cell-array test. The SDD open-circuit test treats an SRAM cell as containing two separate inverters and testing each of them "independently" with "0" and "1" inputs for proper switching response at the output. Figure 5.7 shows an SRAM cell as two separate inverters, the left inverter containing transistors $Q1$ and $Q3$ and the right inverter containing transistors $Q2$ and $Q4$.

Note that the input of the inverter is connected to one of the bit lines, while the output of the inverter is connected to a complementary bit line. Therefore, if we isolate one of the two inverters such that it solely charges/discharges the bit line capacitance connected to its output, then any open-circuit defects or grossly abnormal transistor characteristics will affect the switching time of such an inverter. Open defects in the drain, channel or source path can result in no switching at the output, while a floating gate can result in a lower drain current leading to slower switching time. The implementation of the SDD technique requires testing each of the inverters independently. This could not be done under the normal operating conditions because of the positive feedback loop formed by the two cross-coupled inverters in an SRAM cell. However, it could be overcome by lowering the array power V_{DD} (e.g., from 5 V to $\simeq$2.9 V) during the SDD test, while maintaining the normal word line voltage V_{WL} (e.g, 5 V). It was discovered that for a given cell design there was a critical array supply voltage V_{DD} below which the inverter of an SRAM cell will not switch when a "1" is applied at the input. The test threshold for passing or failing the SDD open-circuit test was set at 150 ns as a reference.

Detecting floating-gate PMOS pull-up transistors with SDD requires a special test mode. Such a transistor could inadvertently pass the SDD test due to the gate charging that may cause a significant drain current. The minimum effective conducting channel current for a floating-gate PMOS transistor to pass the SDD open-circuit test was determined to be much higher than the normal cell-array leakage current. Therefore, the SDD could detect a floating-gate PMOS pull-up by setting the threshold of the cell array leakage current test between the typical array leakage and the minimum "on" current of a PMOS transistor with a floating gate. The NMOS pull-down transistor is reported to be testable by this technique as well. During the SDD test, the bit line connected to the output of the inverter under test was precharged to the level opposite to the final expected level.

In order to detect all possible "soft" defects in an SRAM cell, the SDD technique included the following tests:

- SDD Open-Circuit Test
  - Right inverter with a "0" input
  - Right inverter with a "1" input
  - Left inverter with a "0" input
  - Left inverter with a "1" input
- Cell-Array Current Test
  - Array current test for all "0" data background
  - Array current test for all "1" data background

The SDD technique requires a number of circuit modifications, some additional column circuitry and voltage level manipulations. Figure 5.7 shows the bit line connected to the input of the inverter under test is driven during the SDD test. Meanwhile, the complementary bit line, which is connected to the output of the inverter under test is monitored. Effectively, simultaneous write and read operations are performed on the CUT. The read and write muxes in the SDD-enabled column circuitry were modified such that it was possible to decouple the bit lines during the write and the read operations. Another addition to the column circuitry was the NMOS transistor for the bit line pre-discharge. Pre-discharge is necessary when testing an expected low-to-high transition on a bit line. To ensure that the bit line during the SDD test is driven only by the tested inverter, the normal sense amplifier must be isolated with both a header PMOS transistor and a footer NMOS transistor. Instead, an inverter with an optimized switching point served as the SDD sense amplifier. A special level converter was implemented to control the array V_{DD} during the SDD test. An internal high-speed current detection circuit was added for the SDD array current test. And finally, to accommodate more flexibility in the testable array sizes, a programmable 2-bit delay counter that provides four possible delay times is built in for SDD switching time test limit adjustment. The programmable counter helps to account for the dependence of the bit line capacitance, which loads the inverter during the transition, on the array size. The reported area overhead of adding the SDD-related DFT circuitry to a 16 Kb array was less than 2%. However, a smaller area overhead can be expected when SDD is integrated into larger SRAM arrays.

5.4.1.2 Weak Write Test Mode

The Weak Write Test Mode (WWTM) by Banik et al. [1, 105] is one of the well-known single test stress active techniques for SRAM cell stability test. The

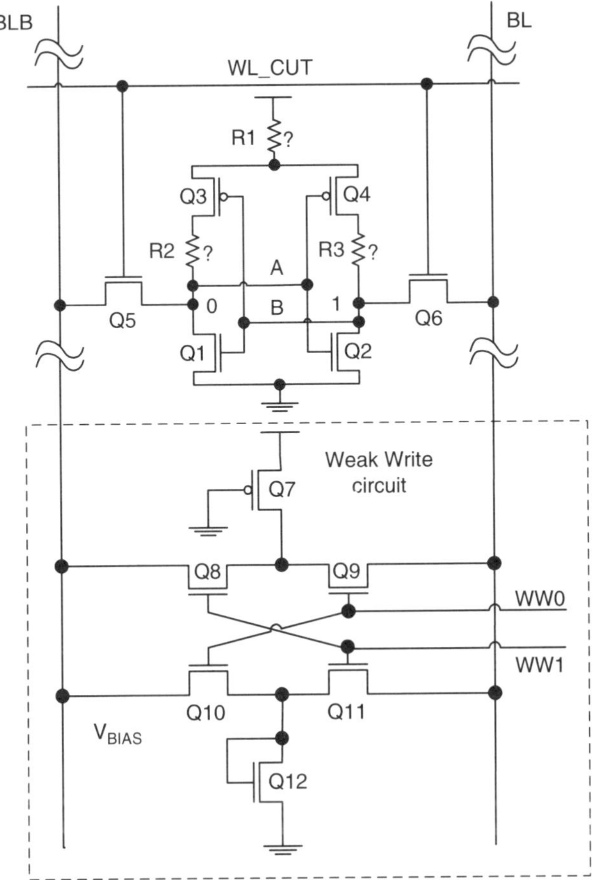

Fig. 5.8 Weak Write Test Mode (WWTM) example circuit [1]

WWTM test requires special DFT circuitry shown in Figure 5.8 and marked as "Weak Write circuit" that is connected to every column of an SRAM array. The DFT circuitry tests the stability of the CUT by attempting to weakly overwrite the CUT. Besides internal defects in PMOS transistors *Q*3 and *Q*4, the WWTM targets resistive symmetric (resistor *R*1) and asymmetric (resistors *R*2 and *R*3) interconnect defects in the pull-up path of the CUT. The WWTM circuit is designed, sized and its weak overwrite current is set such that it is capable to overwrite only a cell with compromised stability that is likely to be caused by a defect in the cell's pull-up current path.

The DFT circuitry can be enabled to perform either a weak write "1" or a weak write "0". The duration of the weak write pulse is typically determined empirically during the design of the weak write circuit. Longer weak write time (about 50 ns) may be necessary to improve the discriminating capability of the WWTM and ensuring that only defective SRAM cells are overwritten during the test [1]. Longer overwrite times are reported to help detect less-resistive defects in the pull-up path of the CUT reaching below 100 kΩ for asymmetric resistive defects. To accommodate several possible weak write times, the access control circuitry should allow the adjustments of the test duration at test time or test duration selection from a preselected set of settings.

When the test pulse is enabled, the pull-down NMOS transistor $Q12$ of the weak write circuit applies a weak overwrite stress on node B of the CUT which is storing a "1" through the series-connected NMOS pass gates $Q11$ and $Q6$ of the weak write circuit and the CUT, respectively. This results in PMOS pull-up transistor $Q4$ contending with the weak overwrite stress current supplied by the pull-down transistor $Q12$ of the weak write circuit. Similarly, the PMOS pull-up transistor of the weak write circuit $Q7$ passes a weak overwrite stress to node A of the CUT which is storing a "0". The NMOS pull-down transistor of the CUT $Q1$ is in contention with the weak overwrite stress supplied through NMOS pass gates $Q8$ and $Q5$ of the weak write circuit and the CUT, respectively. When the weak write "1" line goes active, transistor $Q12$ defines the weak overwrite current applied through transistors $Q11$ and $Q6$ to node B. Meanwhile, transistor $Q7$ prevents the complementary bit line BLB from falling below a voltage level of $V_{DD} - V_{TH\_Q8}$. Alternatively, when the weak write "0" line goes active, the weak overwrite current is defined by transistor $Q12$ and is passed to node A by pass transistors $Q10$ and $Q5$. And transistor $Q7$ maintains the bit line voltage at the level of at least $V_{DD} - V_{TH\_Q9}$.

When the CUT contains an asymmetric defect (e.g., $R3$) and transistor $Q4$ of the CUT is defective, then $Q4$ cannot provide sufficient pull-up current. During the weak write test, the pull-down NMOS transistor $Q12$ of the weak overwrite circuit overcomes the pull-up PMOS transistor $Q4$ of the CUT. The voltage on node B of the CUT is driven down by the reducing voltage of the bit line BL below the typical "1" value voltage. Concurrently, since the gate of the pull-down transistor $Q1$ is connected to node B, the declining voltage on node B reduces the current drive of transistor $Q1$. At the same time, the pull-up transistor of the weak overwrite circuit $Q7$ drives node A up through pass gates $Q8$ and $Q5$. The voltage on node A, which stores a logic "0" is thus driven above the normal level. As voltage on node A reaches the threshold voltage of NMOS transistor $Q2$, it turns on and further degrades the "1" value voltage stored in node B. Since transistor $Q4$ is not operating effectively, the voltage on node B declines to the point where the positive feedback mechanism flips the defective CUT. Similarly, the other data background can be tested.

If the CUT contains a symmetric defect in the pull-up path ($R1$), such as a defect that affects both the inverters $Q1, Q3, Q5$ and $Q2, Q4, Q6$, then both data backgrounds are likely to fail the weak write test. This property can differente between a symmetric and an asymmetric defect in the CUT.

Since during the WWTM test the test circuitry is directly connected through the bit lines of one of the columns to the CUT (through the column muxes), the relative sizing of the pull-up and pull-down transistors in the DFT circuit must be carefully characterized to ensure the the test will not overwrite the cells that are sufficiently stable and are deemed good. Sizing of transistors $Q7$ and $Q12$ is a function of the

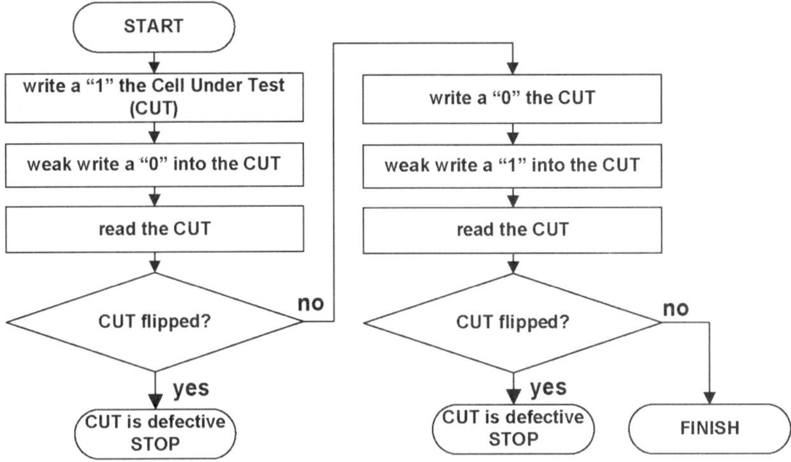

Fig. 5.9 Weak Write Test Mode (WWTM) flow chart [105]

size of the SRAM array, the size of the pass gates in the weak overwrite circuit and the CUT ($Q8 - Q11$ and $Q5, Q6$, respectively) and the transistor sizing of the cross-coupled inverters of the CUT ($Q1 - Q4$) [105].

The generalized test flow is shown in Figure 5.9. The WWTM test starts with writing a "1" into the CUT. Each of the steps can be applied row-by-row to the entire SRAM array. The first step is followed by enabling the DFT circuitry that attempts to weakly overwrite the "1" stored in the CUT with a "0". Once the weak overwrite phase of the test is done, the DFT circuitry is disabled and the array is read normally to determine if any cell has been overwritten. In case a CUT was overwritten, it is deemed defective and the test stops. Otherwise, the test continues with the opposite data background written into the cells. It is followed by the DFT circuitry trying to weakly overwrite it with the complementary value. The subsequent normal read operation determines if any CUT has flipped. In case a CUT has been overwritten by the complementary value, it is deemed defective and the test stops. Otherwise, the Weak Write Test Mode test is considered successfully passed.

The WWTM can replace both the Data Retention test and the Read Disturb test, while requiring a significantly shorter test time. Unlike the passive DRT and the Read Disturb tests that rely upon leakage of NMOS transistors to be effective and thus are more effective at higher temperatures, the active WWTM does not need elevated temperatures to be effective. The WWTM is reported to be more efficient at normal temperatures due to higher transistor currents and thus, a more stable voltage created by the voltage divider in the DFT circuitry. Application of this stability test enables us to distinguish symmetric and asymmetric defects in the pull-up path of the CUT, while a characterization of the weak write time and V_{DD} sensitivity can indicate the severity of the defect.

5.4.1.3 Integrated Weak Write Test Mode

Weiss et al. proposed an SRAM cell stability test technique using a weak write test circuitry integrated into the write driver [106]. Since a write drive is already a part of the column circuitry in the normal operation, using it for cell stability test is advantageous from the area standpoint. Integrating the test circuitry with the write driver allows reduction in the area overhead from the additional transistors which are used only during the test mode.

A typical write driver is designed to quickly discharge the bit line to the ground potential so that any cell within the statistical distribution of the array will flip its state. If a write driver is modified such that the discharge current provided to the bit line is limited to a controlled value, a weak write mode can be achieved with the write driver. Thus, such a modification to the write driver can allow stability testing of the CUT.

One implementation of the WWTM circuit which is integrated into the write driver is shown in Figure 5.10 [106]. The input signal to be written into the cell is

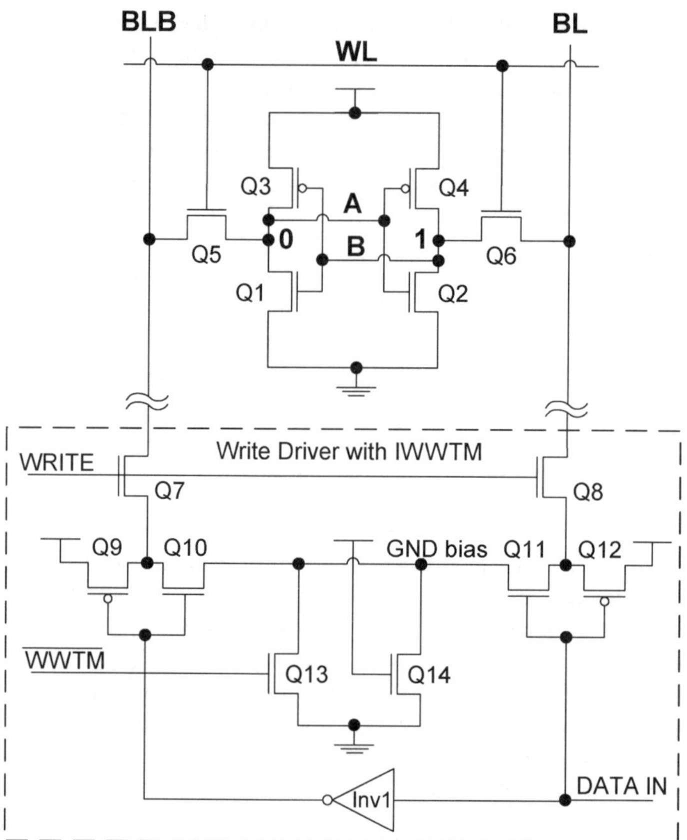

Fig. 5.10 Integrated Weak Write Test Mode (IWWTM) example circuit [106]

supplied to the input of inverter $Q11-Q12$, while its complementary value is provided by the inverter $Inv1$ to the input of inverter $Q9-Q10$. Once $WRITE$ signal is enabled, the data to be written to the cell are passed onto the bit lines through the pass gates $Q7$ and $Q8$. If the sources of transistors $Q10$ and $Q11$ were connected directly to the ground, this arrangement would represent a typical write driver. However, rather than being directly connected to the ground, the sources of transistors $Q10$ and $Q11$ are connected to the $GND\ bias$ node. The $GND\ bias$ node is separated from the ground potential by transistors $Q13$ and $Q14$. Transistor $Q13$ is designed to be strong enough to provide a solid ground potential during a normal write operation when $\overline{WWTM}$ signal is high. The gate of transistor $Q14$ is tied to V_{DD}. This transistor is sized much smaller than $Q13$ such that to limit the write current to such a degree that only cells with weak pull-up current will be overwritten. Thus, in the stability test mode, i.e., when transistor $Q13$ is turned off, transistor $Q14$ effectively converts a typical write driver into a write driver with an Integrated Weak Write Test Mode (IWWTM). The sizing of transistor $Q14$ will vary depending on the circuit parameters of the particular write driver and the associated memory array.

5.4.1.4 Word Line Driver Underdrive During a Write Operation

Schwartz proposed a weak overwrite stress to the CUT by reducing the supply voltage of the word line driver [107]. The technique was coined "Data Retention Weak Write" test (DRWW). Figure 5.11 explains the operating principle of this cell stability test technique.

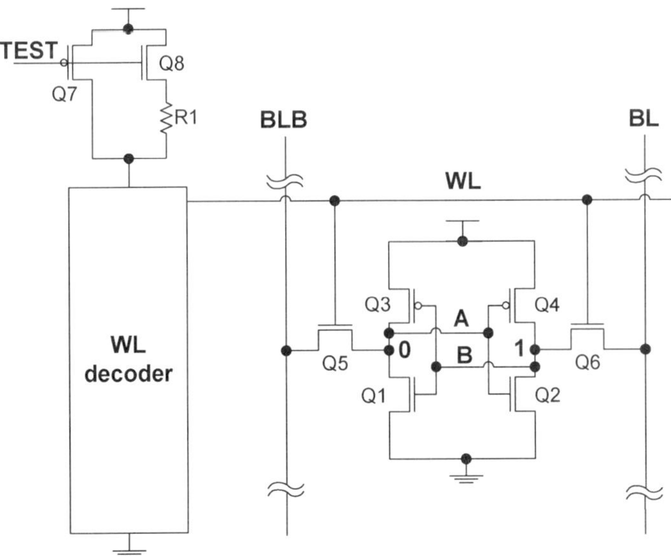

Fig. 5.11 Integrated Weak Write Test Mode (IWWTM) example circuit [107]

The power supply connection of the word line driver is connected to V_{DD} through serially connected NMOS transistor $Q8$ and resistor $R1$ (Figure 5.11). In parallel with the serially connected $Q8$ and $R1$ is a PMOS transistor $Q7$. The gates of $Q7$ and $Q8$ are connected to a common test line "$TEST$". In the normal operating mode, the test line "$TEST$" is set to a logic "0". Thus, transistor $Q7$ is turned on, while transistor $Q8$ is turned off. Transistor $Q7$ is sized such that the word line driver is supplied with full V_{DD} voltage and the word line drivers are provided with enough current to overwrite any cell in the SRAM array under test.

In test mode, the test line "$TEST$" is set to a logic "1". Therefore, transistor $Q7$ is turned off, while transistor $Q8$ is turned on. The word line driver is supplied with a voltage around $(V_{DD} - V_{THn})$. Resistor $R1$ may be needed for setting the desired supply voltage more accurately. Transistor $Q8$ and resistor $R1$ are designed such that the word lines are set to a voltage, which is slightly lower than the threshold voltage of access transistors in the memory array. It was reported that in a typical 3.3 V technology, the write current of 200 μA is sufficient to overwrite any of the SRAM cells in the entire cell population of the array. Schwartz identified that one fourth of the typical write current, i.e., a write current of 50 μA is a good measure to set the write current for a weak write test. Cells that change states after the application of a weak overwrite stress provided by a write current limited to 50 μA are deemed weak/defective. Whereas, the good SRAM cells should withstand the weak overwrite stress.

The test algorithm proposed by Schwartz includes writing a background data pattern in normal mode followed by a read-back verification step. Next, the DRWW circuit is put in the test mode and an inverse data pattern is written to the array using a weak write cycle where the voltage supply to the word line drivers is sufficiently lowered so that the inverse data is inefficiently coupled into the cells under test. Once the weak write cycle is finished, the array is read to verify if the previously written data are still intact. This process is then repeated using inverted data for the initial write cycle, followed by a weak write cycle with the corresponding inverse data.

5.4.1.5 Soft Defect Detection-II (SDD-II) Technique

SRAM cell instability is often caused by subtle defects. While detection of catastrophic defects in SRAMs can be achieved with traditional test methods, weak or "soft" defect detection requires application of special detection techniques. Therefore, a new approach to the Soft Defect Detection (SDD) was taken by Motorola [108]. To distinguish it from the first SDD technique proposed by Kuo et al. [3] and described in Section 5.4.1.1, we will call this test technique SDD-II.

The SDD-II flow chart is shown in Figure 5.12. Soft Defect detection starts with writing of a test state into the SRAM array under test. Next, after a normal precharge of the bit lines, the bit line precharge circuits are disabled that leaves the bit lines floating. In the next step, the floating bit lines are shorted. This manipulation with the bit lines preconditions the bit line potentials for the next step. Subsequently, the

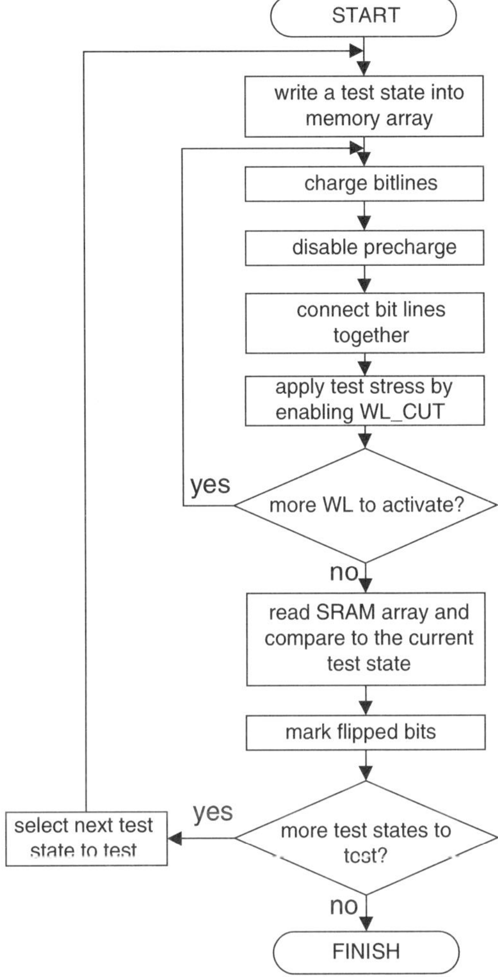

Fig. 5.12 Soft Defect Detection-II (SDD-II) flow chart [108]

word line of the cell under test is activated. The potential of the preconditioned bit lines is applied to the CUT, thus stressing the CUT. If there are more word lines to test, i.e., there are more cells to test with the SDD-II technique, then the bit line preconditioning followed by the word line application is repeated for the other cells in the SRAM array. After all cells in the array are stressed with SDD-II, a read operation is performed on every SRAM cell to determine whether or not they have flipped their state. The cells which flipped their state as a result of SDD-II test are marked as weak. If there are more test states to test (such as with the CUT in the complementary state), the SDD-II test sequence is repeated from the beginning. The necessary transitions between the steps of SDD-II test technique may be implemented using a state machine. Once all cells and their states are tested, the SDD-II test is complete.

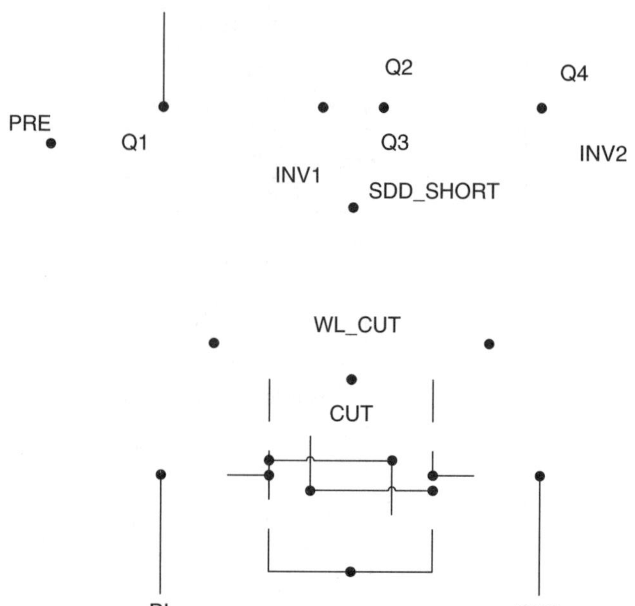

Fig. 5.13 Soft Defect Detection-II (SDD-II) example circuit [108]

An example of a SDD-II circuit is presented in Figure 5.13. The SDD-II circuit uses a discharging transistor $Q1$ and precharging transistor $Q4$ and a transmission gate composed of transistors $Q2$ and $Q3$. SDD precharge signal PRE is applied to the gate of transistor $Q1$ and the complementary version of the precharge signal is applied to the gate of PMOS transistor $Q4$ through inverter $INV2$. The transmission gate is controlled by $SDD\ SHORT$ signal, which is connected to the gate of transistor $Q3$ and through inverter $INV1$ to the gate of transistor $Q2$.

The SDD-II test circuitry preconditions the bit line to redistribute the bit line potentials before enabling the word line of the CUT. Upon application of the SDD precharge signal PRE, the bit line BL is discharged to the VSS potential. Whereas the complementary bit line BLB is precharged to VDD. In the next step of SDD-II stability test, the bit lines are shorted by enabling the $SDD\ SHORT$ signal. Provided that the bit line BL and the complementary bit line BLB have identical capacitance, the charge stored on the bit line BL and the complementary bit line BLB will be shared equally. The resulting bit line voltage $V = q/C$ in this implementation of the SDD-II technique will be near $VDD/2$. When the word line of the CUT is enabled, the the CUT is connected to the bit line BL and the complementary bit line BLB while $V_{BL} = V_{BLB}$. This puts all the transistors of the CUT are in conductive state. If there are no defects in the CUT, conditioning of the bit lines and activation of the $WL\ CUT$ does not result in the state change of the stored value. The storage nodes of a defect-free SRAM cell will be disturbed by the test stress created by the SDD-II. The voltage of the node storing a "0" will rise to a higher level, whereas the potential of the node storing a "1" will dip. However, a stable SRAM cell can withstand such a disturbance and keep the stored data. In case the CUT contains a defect, the stability of such a cell will be impaired. For instance, if the pull-up PMOS transistor of the cell node storing a "1" is open or highly resistive, then connecting such a CUT to the bit lines which are preconditioned using SDD-II will lead to destroying the stored data. The flipped cells will be detected by the read operation conducted on the whole SRAM array after all cells in the array have been stressed with SDD-II test technique.

The change of the stored state in the CUT occurs upon activation of the corresponding word line, which allows stability testing at speeds substantially similar to the normal operation speed of the SRAM. Therefore, a long pause such as used in DRT is not needed because the state change (assuming a defect exists) occurs immediately in response to the word line activation. Note that the SDD-II test is done on multiple SRAM cells in parallel. All bit cells along a selected word line may be conditioned in parallel for subsequent word line activation. In this manner, activation of a single word line affects multiple bit cells which further reduces the testing time.

5.4.1.6 No Write Recovery Test Mode

Yang et al. proposed a No Write Recovery Test Mode (NWRTM) [109], which is capable of replacing the DRT and detecting all defects related to inadequate pull-up current in an SRAM cell. The NWRTM includes a special preconditioning of the bit lines for applying the test stress. A special write cycle is created to precondition the bit lines is followed by a normal write operation.

Figure 5.14 describes the concept of the NWRTM by comparing a normal write operation with a "No Write Recovery" (NWR) write cycle. Suppose that we are trying to write a "1" to node A. During a normal write operation shown in Figure 5.14(a), before the write operation begins, both the bit line BL and the complementary bit line BLB are typically precharged to a full V_{DD} potential, so that $V_{BL} = V_{BLB} = V_{DD}$. After the bit line precharge, the word line WL is activated and the write driver, which is represented by transistor $Q7$ in Figure 5.14(a), is turned on. The write driver current I_{Q7} discharges C_{BLB} from V_{DD} to 0 V. Since $WL = V_{DD}$, the access transistors $Q5$ and $Q6$ are on. Transistor $Q6$ pulls down node B that stores a "1". Once the gate of transistor $Q3$ is driven below $V_{DD} - V_{TH\_P}$, transistor $Q3$ creates a pull-up current I_{Q3} that pulls node A up. As soon as the potential on node A exceeds V_{TH_n} of transistor $Q2$, the positive feedback of the latch mechanism of the SRAM cell flips the cell data concluding a normal write operation.

To apply the NWRTM, a special write operation, i.e, a NWR, must be performed before a complementary normal write operation tests the stability of the CUT. To illustrate the concept of the NWRTM, consider an SRAM cell with defect resistance $R1$ in the pull-up path of node A shown in Figure 5.14(b). The difference between a normal write operation and the NWR is that unlike the normal write, in the NWR write cycle bit lines are not precharged back to V_{DD}. Instead, the bit lines

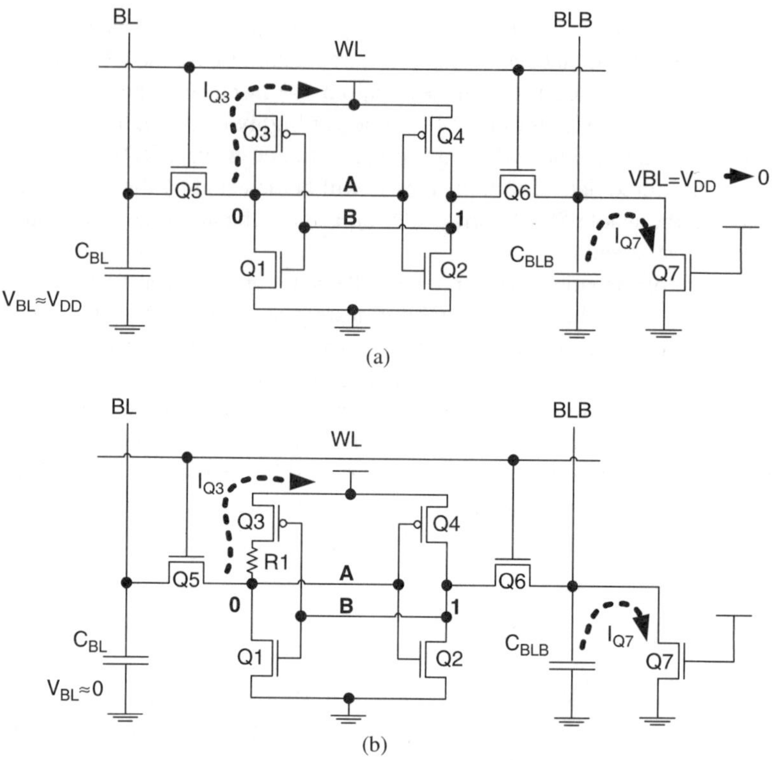

Fig. 5.14 (a) Defect-free SRAM cell during a normal write operation; (b) The concept of detecting a defect with resistance $R1$ using the No Write Recovery Test Mode (NWRTM)

are left floating and remain in the previous state, i.e., the bit lines are not "recovered". For instance, if a "0" was written to node A using a NWR write cycle, then after the NWR write cycle completes, the bit line voltage V_{BL} will remain near 0 V (Figure 5.14(b)). Therefore, when a "0" is subsequently written to node B, the only path to charge node A is through the pull-up transistor $Q3$. The potentials of the bit line V_{BL} and the complementary bit line V_{BLB} are said to be set to a weak GND and a strong GND, respectively. If for any reason, such as the presence of a defect with resistance $R1$ and/or the pull-up current I_{Q3} is not sufficient, then node A will not be pulled up high enough for the cell to be overwritten. In other words, a defective SRAM cell will not be overwritten as a result of applying the NWRTM while a good cell will flip the states and will be overwritten.

Note that the behavior of a defective cell tested with the NWRTM is opposite of that tested by the majority of other DFTs designed for SRAM cell stability test. While other techniques detect an unstable SRAM cell by flipping the states of a defective cell, the NWRTM flips only a good cell while a defective cell retains its state. A subsequent read operation determines whether or not the CUT flipped as a result of the NWRTM. To complete the test of the CUT, the NWRTM is then performed with complementary data written to the CUT follower by a verifying read operation.

An example of march test elements that adds the NWRTM functionality to the standard march test set is shown below:

$$\Downarrow (w0, NWR1, r1) \Uparrow (w1, NWR0, r0)$$

In the first step of the first element, the CUT is normally written a "0". Then, the bit line precharge is disabled and a "1" is written in a NWR1 step. Next, a read operation with an expected value of a "1" is performed to verify whether or not the CUT had flipped. In the second step, the same march operations are performed with complementary data applied to the CUT.

Implementation of the NWRTM requires a special test mode circuitry that controls the enabling/disabling timing of the bit line precharge transistors in SRAM array. The advantage of the NWRTM is a minimal area overhead in the SRAM array and no impact on the array performance.

5.4.1.7 Word Line Voltage Overdrive During a Read Operation

It is well known that a read access to an SRAM cell creates a disturbance on the storage nodes. Since bit lines in the majority of modern SRAM designs are precharged to full V_{DD}, it is the node storing a logical "0" that will experience the disturbance during a read access to the cell. Practically, the voltage level on the "0" node can easily increase from 0 V in the standby state to up to 20% of V_{DD}. Ideally, a properly designed and fabricated defect-free SRAM cell should withstand a read access disturbance without loosing the stored data value. In practice, the growing process variations and imperfections in deep sub-micron technologies can cause a number marginally stable cells in a large array. It is a difficult task to expose such unstable cells using regular memory test methods. To address this problem, Salters proposed a stronger read disturbance to test the stability of SRAM cells and expose marginal cells in the array [110]. The stronger read disturbance is achieved by applying a higher supply voltage to the row decoder and hence, to the WL of the CUT. The schematic representation of this testing concept is shown in Figure 5.15.

The power supply of the word line drivers in this scheme requires the capability to switch from the regular operating V_{DD} value to an elevated value V_{DDhigh}. The elevated voltage (overdrive) V_{DDhigh} can either be supplied by an embedded charge pump or by an external power pin for the word line driver supply. When using an external pin, it can be controlled by the tester during the test and then be connected to V_{DD} after the testing is finished. Alternatively, a similar disturbance effect can be achieved by lowering the supply voltage of SRAM array while keeping the supply voltage of the write drivers at V_{DD}, or by a combination of both. The stability test using word line overdrive contains the following steps. First, a data background is written into the SRAM array by a normal write operation. In the next step, a higher supply voltage V_{DDhigh} is applied to the word line drivers. As a result of this overdrive, the access transistors $Q5, Q6$ of the CUT (Figure 5.15) become stronger than in a normal mode. This, effectively, changes the cell ratio of the CUT and makes it more susceptible to a read disturbance. The overdrive voltage is chosen such that if

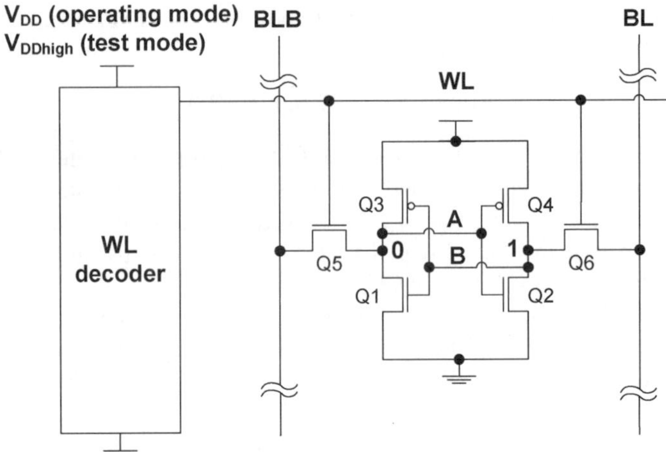

Fig. 5.15 SRAM cell stability test by elevating the word line voltage [110]

the CUT has a sufficient SNM, it will withstand this disturbance. However, if the noise margin of the CUT for any reason is compromised beyond a pre-determined threshold, such a cell will flip and loose its data value. A subsequent normal read operation will detect such unstable cells. If no cell had failed with the first data background, an opposite data background is written into the array and the read disturbance test followed by a verification read step are repeated.

This test method does not require additional circuitry put into the memory array and thus has minimal effect on the array area. The area penalty, however, is introduced by the charge pump that is required to provide the overdrive voltage to the word line drivers. If this technique is implemented using an extra power supply pin, it may put additional strain on pad-limited designs.

Many of the single detection threshold methods can be altered to enable analog control of the weak overwrite stress. However, global analog levels are more difficult to control if implemented internally or more pad- and tester-demanding if controlled at the ATE level.

5.4.2 Active Techniques with Programmable Test Stress Levels

Thanks to the programmability of the test stress, the amount of test stress can be determined/adjusted after the design stage is complete. The SRAM cell stability test techniques with a single test stress attempt to apply the preselected proper test stress across wide ranges of process variations and cell defect mechanisms. In contrast, the programmable test techniques offer lower design effort and a reduced risk of applying a non-optimal test stress. In fact, the lack of flexibility in setting the amount of test stress is a major disadvantage of SRAM cell stability test techniques with a single test stress level. A non-optimal amount of stress, in turn, will result in a non-optimal detection threshold for the cells that are deemed insufficiently stable.

One possible reason why the single-stress SRAM cell stability test techniques may produce non-optimal test stress in deep sub-micron technologies is the growing difficulty of accurate process modeling. Inaccurate process models can cause sub-optimal sizing of the test circuitry as well as the unexpected stability characteristics of SRAM cells under test. Another reason calling for SRAM cell stability test with programmable test stress level is the growing process complexity and increasing process variations and changes. Scaling transistor and layout features in modern technologies introduces multiple challenges for process developers which are the main reason for appearing of programmable stability test techniques.

Inherently, the variability of transistor threshold voltages, I_{ON} and I_{OFF} and other parameters is becoming more difficult to control with every new technology generation. Therefore, allowing the test stress to be controlled during the production process helps to mitigate the risk of the fabricated SRAM cells not operating exactly as it was intended during the design stage. Such a risk may be a result of using incomplete or inaccurate process models during the design stage, while the manufactured SRAM array could exhibit a different or unexpected behavior.

Programmability of the test stress enables balancing the number of test escapees due to insufficient test stress and the test yield loss caused by the excessively stringent stability testing that fails some of the good SRAM cells. Choosing the correct stress for SRAM cell stability testing non-trivial. Making the complex decision of drawing a line between good stable cells and defective unstable cells is often possible only after analyzing the test characterization results of a significant population of SRAM cells in high-volume manufacturing.

5.4.2.1 Short Write Test Mode

A programmable test technique for SRAM stability test, the Short Write Test Mode (SWTM), was proposed by Mehalel [111]. The method is referred to as a "short write" test because it uses a special write operation which is short enough not to cause a robust SRAM cell to flip. The strength of the short write drive circuit is designed such that if it were applied for a sufficiently long period, it would be sufficient to trip even a stable SRAM cell. A weak SRAM cell, however, would flip after application of even a short write pulse, therefore indicating its defective status. The SWTM bears all the benefits of an active programmable stability test technique: the ability to replace the lengthy delays of the pause (DRT) test while providing fine tuning of the applied test stress on the fly to cover a wide spectrum of process parameters and the target defects.

An example of incorporation of the SWTM into a march test is shown below:

$$\Downarrow (w0, SW1, r0) \Uparrow (w1, SW0, r1).$$

In the first step of the first march element, the CUT is normally written a "0". Then, SWTM circuitry attempts to overwrite the CUT with a short write of a "1". This is followed by a verifying read operation with an expected value of "0". The same test operations are next performed with complementary data values applied to the CUT.

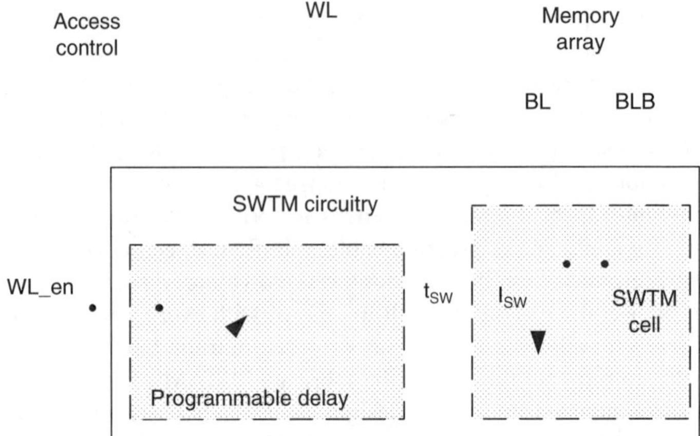

Fig. 5.16 Short Write Test Mode (SWTM) concept [111]

A block diagram explaining the concept of the SWTM is shown in Figure 5.16. Figure 5.16 depicts one of the two possible short write tests. The SWTM cell applies a short write current I_{SW} to the bit line (BL) and holds the complementary bit line (BLB) at the supply voltage. To realize a short write with a complementary data, the short write current I_{SW} is applied to the complementary bit line (BLB) while the bit line BL is held at the supply voltage. A word line enable (WL_{en}) signal controls the programmable delay circuit to generate a short write test pulse t_{SW} of a controlled width that activates the short write driver connected to a bit line.

Figure 5.17 shows a possible circuit implementation of the SWTM. The left part of Figure 5.17 presents the Software-Controlled Timing (SCT). The SCT block is used for the pulsewidth programming of the short write pulse. The short write pulse is applied to the SWTM circuit. The SWTM circuit contains short write driver transistors $Q10$ and $Q12$, and pass transistors $Q9$ and $Q11$. Transistor pairs $Q9-Q10$ and $Q11-Q12$ form short write current sources that generate I_{SW}. Transistors $Q7$ and $Q8$ serve as test precharge drivers and are controlled by the designated test control circuitry.

The software-controlled timing block uses a number of individual SCT memory elements $SCT\ 0$ to $SCT\ N$. Each of the SCT elements controls a pass gate $Q\ PASS$ which is connected to a pull-down transistor $Q\ PD$. The pull-down transistor gates are controlled by the wordline enabling signal WL_{en}. The short write pulse is generated by an AND gate $AND1$ and buffers $B1$ and $B2$ in conjunction with the elements of the SCT array. The pulsewidth of the short write pulse t_{SW} has two portions - a programmable portion t_{PRG} and a constant portion t_C:

$$t_{SW} = t_{PRG} + t_C.$$

The programmable portion t_{PRG} the short write pulse t_{SW} can be programmed by storing different values in memory elements $SCT\ 0$ to $SCT\ N$. When the pass transistors Q_{PASS} are enabled, more pull-down transistors are connected in parallel

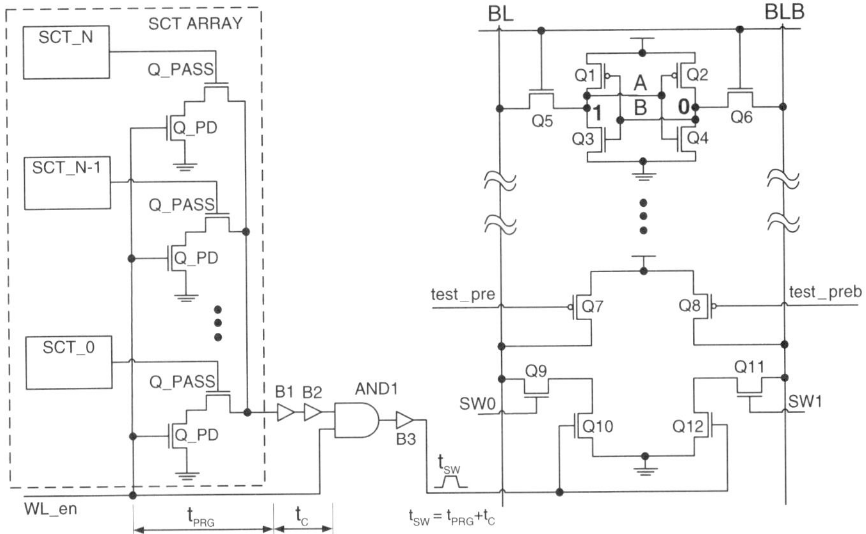

Fig. 5.17 Short Write Test Mode (SWTM) circuit implementation [111]

Table 5.1 SWTM control signal states

Mode	Q7 gate	Q8 gate	Q9 gate	Q11 gate	Q10/Q12 gates
Short write 0	1	0	1	0	Pulse
Short write 1	0	1	0	1	Pulse
Normal operation	1	1	0	0	0

therefore speeding the transition of the short write pulse by shortening its programmable portion. The constant portion t_C is defined by the total delay of buffers $B1$ and $B2$.

Table 5.1 gives the SWTM logical states in the two short write test modes (short write of a "1", short write of a "0") and in normal operation:

The SWTM cell which is connected to each column of an SRAM array can use a regular SRAM cell with modified connections. Transistors *Q_PASS* and *Q_PD* can also be sized identically to the current sources in the SWTM cell. Equally sized, these transistors add the benefit of a better tracking of process changes as various current sources will likely be affected similarly by the varying process parameters. The short write control signals are inactive during normal read and write cycles. The additional capacitance added to the bit line capacitance by the SWTM test circuitry is considered negligible in most larger SRAM arrays.

Figure 5.18 shows the waveforms when the SWTM test circuitry is driving the bit line BL low by the short write pulse with a duration t_{SW}. The short write voltage V_{SW} is set by programming the SCT memory array and is a function of I_{SW} and the bit line capacitance C_{BL}. The CUT in Figure 5.18 recovers from application of the short write voltage V_{SW}. Whereas a defective cell would fail to return to its original

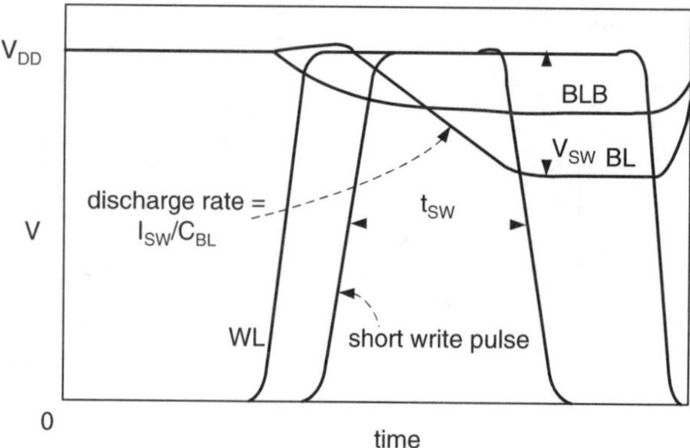

Fig. 5.18 Short Write Test Mode (SWTM) example waveforms [111]

state. Assuming that the memory cell and the current sources in the SWTM cell are identically sized, the approximate value of V_{SW} can be expressed as:

$$V_{SW} = \frac{I_{SW} \times t_{SW}}{C_{BL}}$$

Assuming the buffers $B1$ and $B2$ trip at $V_{DD}/2$, the programmable portion of the short write pulse can be approximated as:

$$t_{PRG} = \frac{(V_{DD}/2) \times C_{BL}}{nI_{SW}}$$

where n is the number of enabled SCT memory cells. Thus, the short write voltage V_{SW} can be broken down into two components:

$$V_{SW} = \frac{V_{DD}}{2n} + \frac{I_{SW} \times t_C}{C_{BL}}$$

The first component can be assumed constant due to identical sizing of the current sources. The effect of process variations on t_C and I_{SW} in the second component is opposite, which is stabilizing the effect of the fixed portion t_C of the short write pulse on V_{SW} voltage across the process variations. Therefore, the programmable portion of the short write pulse defines the desired cell stability test voltage V_{SW}. The desired stability detection threshold can be adjusted based on high-volume manufacturing characterization without the need for additional design iterations.

5.4.2.2 Programmable Weak Write Test Mode

A programmable version of the Weak Write Test Mode (PWWTM) was proposed by Selvin et al. [112]. The original WWTM which was proposed by Banik et al. [105] used on-die test circuitry to stress each SRAM cell with a pre-designed "weak write" stress that will pass a good cell and fail a cell with significant device or interconnect defects. However, the PWWTM allows for programming of the test stress on the fly during the test. Therefore, the PWWTM helps to optimize the test stress on the CUT and to account for the post-silicon factors on the circuit behavior.

The concept of the PWWTM can be clarified using Figure 5.19. The PWWTM example circuit in Figure 5.19 has two parts. The left part is the programmable bias

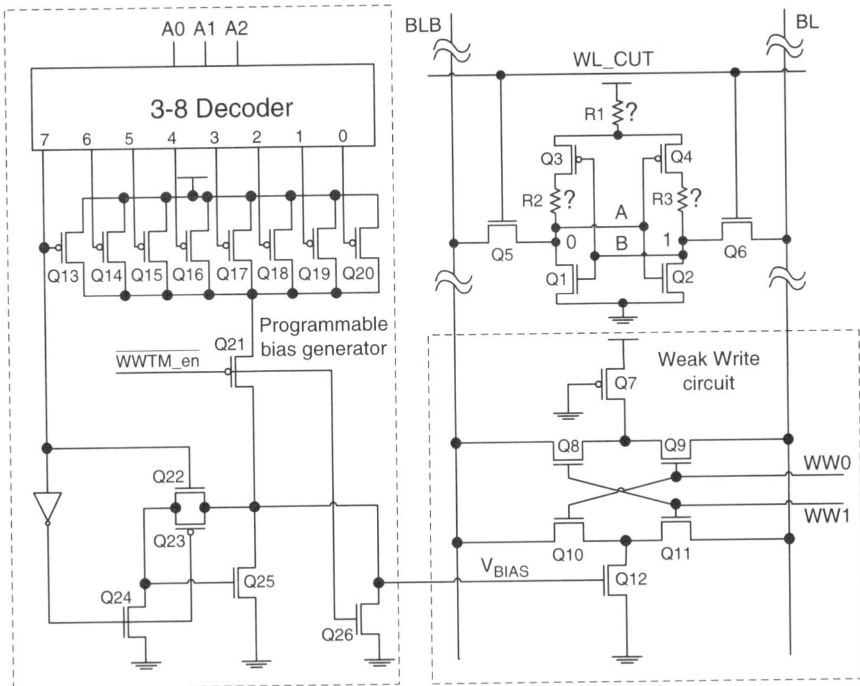

Fig. 5.19 Programmable Weak Write Test Mode (WWTM) example circuit [112]

generator and the right part is a schematic representation of an SRAM column with WWTM circuit connected to the bit lines. One may notice a close resemblance of the right part of Figures 5.8 and 5.19. However, while in the WWTM circuit implementation the gate of NMOS transistor $Q12$ is connected to its drain, the gate of the same transistor $Q12$ in the programmable version of the WWTM is connected to the programmable bias generator shown in the left part of Figure 5.19. In the WWTM circuit, transistor $Q12$ functions as a fixed current source that defines the weak write current. The PWWTM achieves programmability of the weak overwrite current by controlling the gate voltage of transistor $Q12$ with a programmable voltage V_{BIAS}.

When the weak write "1" line goes active, transistor $Q12$ defines the weak overwrite current applied through transistors $Q11$ and $Q6$ to node B. Meanwhile, transistor $Q7$ prevents the complementary bit line *BLB* from falling below a voltage level of $V_{DD} - V_{TH\_Q8}$. Alternatively, when the weak write "0" line goes active, the weak overwrite current is defined by transistor $Q12$ and is passed to node A by pass transistors $Q10$ and $Q5$. And transistor $Q7$ maintains the bit line voltage at the level of at least $V_{DD} - V_{TH\_Q9}$. The drain current of transistor $Q12$ and thus the programmed weak write current is defined by the biasing voltage V_{BIAS} generated by the programmable bias generator.

Assuming that the test mechanism of the PWWTM with V_{BIAS} set to a certain level is similar to that of the WWTM (Section 5.4.1.2), we will concentrate on how the programmable bias generator shown in the left part of Figure 5.19 generates the bias voltage V_{BIAS}. The programmable bias generator is implemented using a programmable voltage divider formed by an NMOS transistor $Q25$ and eight PMOS transistors $Q13-Q20$. Gates of transistors $Q13-Q20$ can be controlled by a 3-to-8 decoder. The bias generator also includes an enabling PMOS transistor $Q21$ controlled by $\overline{WWTM\_en}$ signal. In this implementation the 3-to-8 decoder allows for seven weak write test stress settings and one special mode. During the PWWTM test when binary inputs $A0-A2$ of the decoder are supplied values from "000" to "110", one of the PMOS transistors $Q14-Q20$ is contending with NMOS transistor $Q12$. The gate and the drain of transistor $Q12$ are shorted by a transmission gate formed by transistors $Q22, Q23$. Transistors $Q14-Q20$ are sized such that the bias voltage V_{BIAS} that is formed on the drain of $Q25$ is turning on transistor $Q12$ enough to weakly overwrite a CUT with a particular resistive defect in the pull-up path. Setting $\overline{WWTM\_en}$ signal to a logic "1" breaks the direct *dc* path in the bias generator and disables the generator.

Matching of transistors $Q25$ and $Q12$ reduces the sensitivity of the programmable bias generator to process skews. For instance, when the process skews V_{THn} to a faster corner, the current drive of transistor $Q25$ will increase, while the generated bias voltage V_{BIAS} will reduce. This will help compensate the effect of the process skew on the generated weak write current.

The programmable bias generator also features a special "V_{DD}" mode. This mode is enabled when decoder inputs $A0-A2$ are set to "111". In this special mode, a "0" is applied to the gate of transistor $Q13$. At the same time, the transmission gate ($Q22, Q23$) disconnects the gate and the drain of transistor $Q25$, while transistor $Q24$ turns on and grounds the gate of transistor $Q25$. That turns off transistor $Q25$ and the bias voltage V_{BIAS} is driven to V_{DD} through transistors $Q21$ and $Q13$. The "V_{DD}" mode may be useful to test the Weak Write circuit itself. Since in the "V_{DD}" mode $V_{BIAS} = V_{DD}$, transistor $Q12$ is fully "on" and the Weak Write circuit is expected to be able to flip the CUT. This ensures that the Weak Write circuit is defect free and should function properly.

Transistor $Q26$ forces the programmable bias generator output voltage V_{BIAS} to a "0" when the PWWTM is disabled by setting $\overline{WWTM\_en}$ signal to a "1". Therefore, transistor $Q26$ prevents the V_{BIAS} net from floating and causing potential leakage conditions when the PWWTM is disabled.

The detection capabilities of the PWWTM are illustrated in Figure 5.20. The programmed bias voltage V_{BIAS} is stepped up by changing the decoder inputs $A0-A2$ from "000" to "111" reaching full V_{DD} for the "111". The voltages on the internal node A (solid line) and node B (dashed line) react to the increase of V_{BIAS} with every increment of $A0-A2$. At some point, which depends on the stability of the CUT translated to its pull-up current I_{PUP}, the CUT flips and the internal nodes of the CUT settle to opposite states. The CUT with a certain resistive defect in the pull-up path is shown flipping in Figure 5.20 when the decoder inputs $A0-A2$ are programmed to "010". To determine a correct setting for $A0-A2$ for a given defect resistance in the pull-up path of the CUT (e.g., 1 MΩ [113]), one can simulate the corresponding defect value for each weak write setting. Lower $A0-A2$

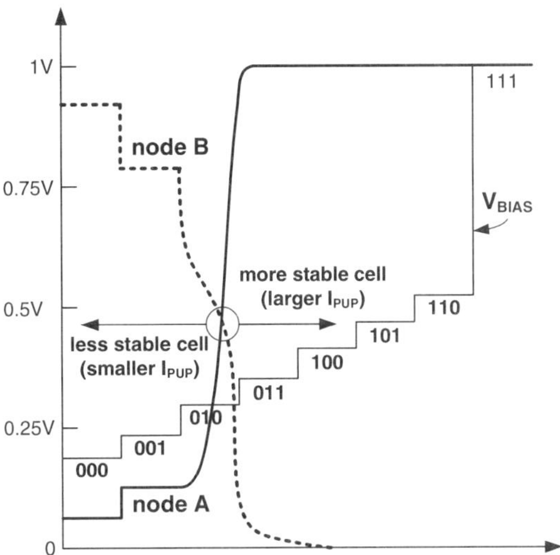

Fig. 5.20 Biasing decoder settings and their effect on V_{BIAS} and the detection capability of the PWWTM [113]

settings will generate a lower V_{BIAS} and thus will create a smaller weak overwrite current. A weaker overwrite current can only detect cells with larger defect resistances. Whereas, the larger V_{BIAS} generated by the higher settings (except for the "111") will create a stronger overwrite current and will detect defects with smaller resistance.

5.4.2.3 Programmable Integrated Weak Write Test Mode

Wuu et al. [114] proposed using a programmable weak write test mode bias generator and combined it with the Integrated WWTM (Section 5.4.1.3) to create a programmable version of the IWWTM. We will refer to this test setup as to Programmable Integrated Weak Write Test Mode or PIWWTM. The PIWWTM uses a programmable bias generator that is similar to the one described in Section 5.4.2.2. However, while the V_{BIAS} of the programmable bias generator discussed in Section 5.4.2.2 equals to "0" when the WWTM is not asserted, the PIWWTM programmable bias generator outputs $V_{BIAS} = V_{DD}$ when the WWTM mode is not active.

Figure 5.21 illustrates the principle of the PIWWTM. Similar to the IWWTM technique described in Section 5.4.1.3, the right part of Figure 5.21 shows a cell under test composed of transistors $Q1 - Q6$. A Write Driver with IWWTM shares the same bit lines with the CUT. In a regular write driver, the sources of transistors $Q10$ and $Q11$ would be connected to the ground (*GND_bias*). However, in a write driver with IWWTM, the ground bias net *GND_bias* is instead connected to the ground through transistors $Q13$ and $Q14$. Transistor $Q13$ is designed to be strong

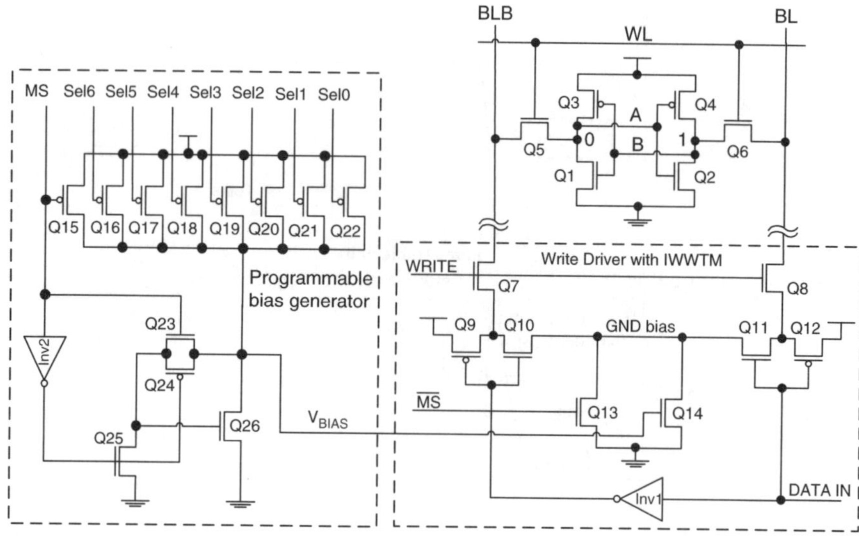

Fig. 5.21 Programmable Integrated Weak Write Test Mode (PIWWTM) example circuit [114]

enough to provide a solid ground potential during a normal write operation when the complementary Mode Select (MS) signal $\overline{MS}$ is high. While in the IWWTM implementation the gate of transistor $Q14$ was tied to V_{DD}, in the programmable IWWTM the gate of transistor $Q14$ is biased by V_{BIAS} supplied by the programmable bias generator shown on the left of Figure 5.21. Thus, in the cell stability test mode, i.e., when transistor $Q13$ is turned off, transistor $Q14$ effectively converts a typical write driver into a weak write driver. The weak write current of the Write Driver with IWWTM can be programmed by manipulating the magnitude of bias voltage V_{BIAS} generated by the programmable bias generator.

As mentioned before, the programmable bias generator implementation is similar to the PWWTM generator and is optimized for the use with the special write driver of the IWWTM. The output bias voltage V_{BIAS} is defined by the ratio of drain currents of transistors $Q26$ and one or a group of transistors $Q16-Q22$. Transistor $Q15$ is controlled by the MS signal. When MS is low, the transmission gate formed by transistors $Q23-Q24$ and inverter $Inv1$ disconnects the gate and the drain of transistor $Q26$, while transistor $Q25$ grounds the gate of transistor $Q26$ ensuring that it is completely turned off. As a result, the output bias voltage V_{BIAS} is driven to V_{DD} through transistor $Q15$. Since during the normal operating mode the bias voltage $V_{BIAS} = \overline{MS} = V_{DD}$, transistors $Q13$ and $Q14$ are both turned on. That improves the connection of the ground bias net *GND_bias* to the ground as opposed to using $V_{BIAS} = 0$ produced by the programmable bias generator in Figure 5.19. To further simplify the PIWWTM bias generator, transistors $Q21$ and $Q26$ shown in Figure 5.19 are not used. Not having a PMOS enable transistor $Q26$ shown in Figure 5.19 reduces the serial resistance in the voltage divider and helps to reduce the transistor sizing and area overhead.

The operation algorithm of the programmable IWWTM is similar to its non-programmable version that was described in Section 5.4.1.3 with the addition of the ability to program the test stress applied to the CUT during the test.

5.4.2.4 Programmable Read Current Ratio Technique with a Pass Transistor Technique

A digital programmable technique using the ratio of read currents of SRAM cells within a column to create a weak write stress to the CUT is described in [6, 115]. It is coined the RCRPT technique.

The concept of programmable threshold is implemented using a set of n SRAM cells in a given column. Either the existing cells in the column or external cells can be used for this purpose. Let R be the ratio of cells having state "0" to the total number of cells in a set n cells (Figure 5.22).

$$R = \frac{number\ of\ cells\ with\ state\ \text{"0"}\ in\ a\ set\ of\ n\ cells}{n}$$

It is assumed that the rest of the cells in a set of n cells have state "1". Initially, both the BL and BLB are precharged to V_{DD}. By manipulating the value of ratio R, and simultaneously accessing the set of n cells, one can manipulate the voltage on the bit lines. For instance, if the number of cells carrying zeroes and ones is equal ($R = 1/2$ in Figure 5.22), then, provided the n cells have the same driving strength, $V_{BLB} = V_{BL}$. Now, suppose that $R > 1/2$. That will cause V_{BLB} to be less than V_{BL}. Respectively, if $R < 1/2$, then V_{BLB} will be higher than V_{BL}.

Now, if a certain ratio of "0"s and "1"s is written to a set of n cells, then the n word lines belonging to the n cells are simultaneously enabled, and then the CUT is accessed, the voltage on cell's nodes (V_{node_A} or V_{node_B}) can be reduced to a given V_{TEST} value (Figure 3.28). When $V_{TEST} > VM_{weak}$, the regenerative property of the cell will restore the stable state and the weak cell will not flip. The target range for V_{TEST} should be set such as $VM_{good} < V_{TEST} < VM_{weak}$. Weak unstable cells will flip and be detected, whereas the cells with satisfactory stability will withstand this stress. This is the selectivity condition of weak cell detection. And finally, when

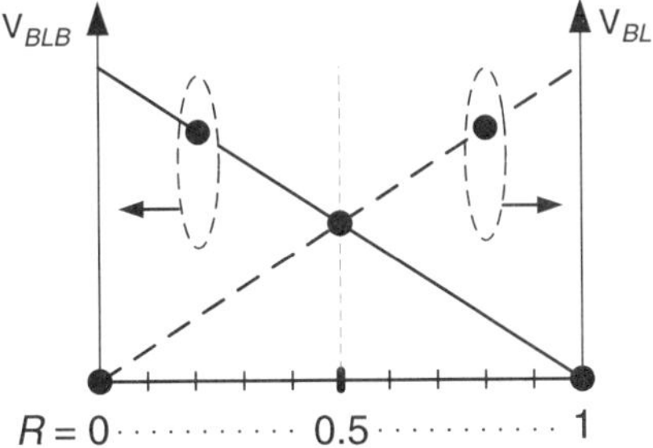

Fig. 5.22 Definition of the programmable ratio R in the proposed detection technique

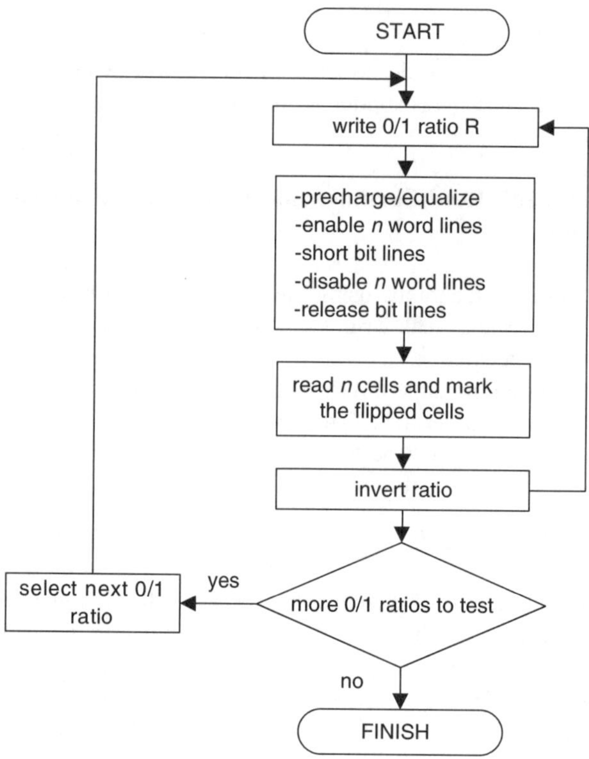

Fig. 5.23 The flow chart of the Programmable Read Current Ratio Technique with a Pass Transistor (RCRPT) Technique [6]

$V_{TEST} < VM_{good}$, even the good cells will flip. Therefore, by varying the ratio R, one can program a detection threshold for detection of SRAM cells with varying degrees of stability degradation.

The flow diagram shown in Figure 5.23 shows the sequence of steps of the RCRPT technique. The cell stability test using the RCRPT technique starts with writing a predetermined 0/1 ratio R into the memory array. Then, the bit lines are precharged and equalized. This is followed by the test sequence, which includes simultaneous enabling of n word lines of the set of the cells chosen to form ratio R. Next, the bit lines are shorted and then the n word lines are released. Once the n word lines are released, the bit lines short is removed. The contents of the SRAM array are then read and compared against the current 0/1 ratio, which was initially written into the array. Any cell whose contents do not match the original 0/1 ratio (i.e., whose state has flipped) are marked or otherwise identified as "weak". In the next step the ratio R is inverted and the test and verification steps are repeated for the inverted 0/1 ratio. The inverted 0/1 ratio is used to detect weak cells that may flip in the opposite direction. These test steps are repeated for all 0/1 ratios that are required to be tested. The resolution of the RCRPT can be refined by increasing of the total number of cells comprising the read current ratio in the set of n cells and/or by changing the pulse width of the pulse enabling the word lines of all n cells.

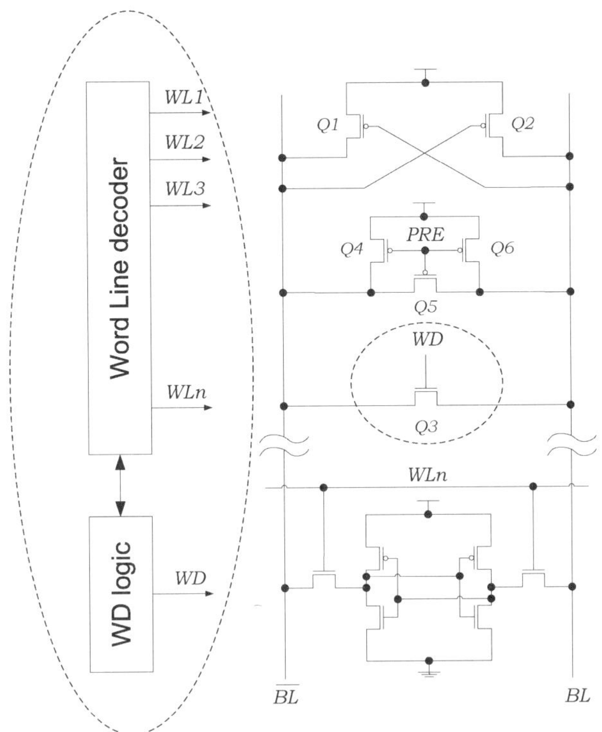

Fig. 5.24 Hardware implementation of the programmable RCRPT, ver. 1 [6]

Figure 5.24 shows the hardware required for one of the implementations of the RCRPT. Ellipses surround additional and/or modified circuitry. Figure 5.24 represents one of the SRAM cells in a column, two cross-coupled PMOS transistors ($Q1$, $Q2$) to pull up the bit lines, three other PMOS transistors ($Q4 - Q6$) to precharge the bit lines to V_{DD} and one NMOS ($Q3$) transistor to short the bit lines. It also includes special logic to issue the Weak Detect (WD) signal, and a modified word line decoder capable of simultaneous enabling of n word lines.

The weak-cell detection phase starts by programming the trip point that is necessary to detect cells with the SNM below the target value. This is done by writing a predetermined number of cells with either a "1" or a "0" state. After the normal bit line precharging finishes, n word lines are simultaneously enabled connecting in parallel n cells of the same column. Under this configuration, access transistors of each side of an SRAM column share a common gate and common bit line nodes. The other terminal of each of the access transistors is connected either to the ground or to V_{DD} through the corresponding driver or load transistors of an SRAM cell. The access transistors work as resistors that are dividing the power supply voltage on each of the bit lines between V_{DD} and the ground depending upon the equivalent dc path resistance. For instance, bit line potentials will be around $V_{DD}/2$ when 50% of cells are in state "0" and 50% of cells are in state "1" because the path resistance to the ground and V_{DD} is the same, i.e. $R = 0.5$ (Figure 5.22).

When the word lines are enabled, the capacitance of each of the bit lines discharges according to the time constant created by the corresponding equivalent path. If the bit lines discharge below the metastable point VM_{good} of the good cells, even the good cells will flip pulling one of the bit lines even further to the ground and restoring the other one to V_{DD}. To prevent the cells from reaching the metastable point VM_{good}, the bit lines are shorted through an NMOS pass transistor by applying a *WD* ("weak detection") pulse. This causes the voltages at the bit lines to remain at around $V_{DD}/2$ while the cell dynamics finds a new equilibrium. The bit lines are not pulled to complementary logical values. However, a bit line voltage around $V_{DD}/2$ is already sufficient to flip the weaker cells (Figure 3.28) with insufficient SNM. For a ratio $R \neq 0.5$, the corresponding path resistances to V_{DD} and the ground will be different and thus the bit line voltage is pulled earlier above or below $V_{DD}/2$.

To prove the effectiveness of the RCRPT test technique a simulation setup with eight six-transistor SRAM cells in CMOS 0.13 μm technology with $V_{DD} = 1.2$ V was used. The degree of weakness of one of the cells was manipulated by varying the resistance value of the resistor between node *A* and node *B* of this cell as per the stability fault model (Figure 3.27). To verify the data retention fault detection capabilities of the RCRPT, the RCRPT was also simulated with intentionally inserted resistive breaks in the source paths of the load PMOS transistors.

Figures 5.25 and 5.26 illustrate the voltage dynamics of node *B* as well as other signals for the weak cell (*int_weak*) and a reference typical cell (*int_typ*). The waveforms are shown for a 0/1 ratios of 3/5 and 5/3 respectively.

The weak cell was forced into a weak state by connecting nodes *A* and *B* with a resistor of 200 kΩ. Evaluation of this cell with the inserted resistor of 200 kΩ gave an SNM of around 50% of the typical SNM for the SRAM cell that was used.

A logical "1" state was stored in node *B* of the weak cell as well as in node *B* of a reference typical cell. After precharging both bit lines to V_{DD} and equalizing them,

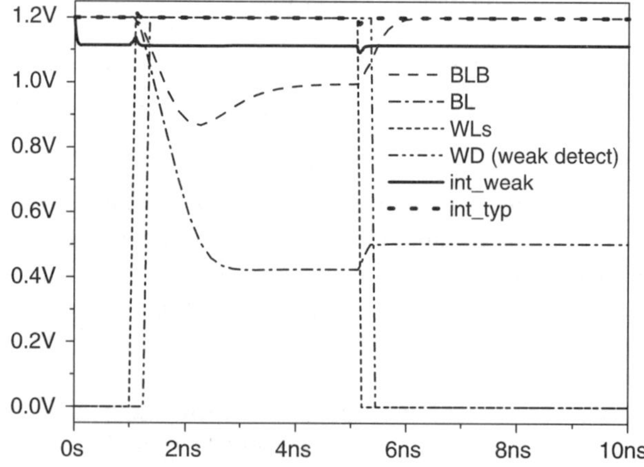

Fig. 5.25 Voltage dynamics of node B and other signals for a weak cell (int_weak) and a reference cell (int_typ) for a 0/1 ratio of 3/5

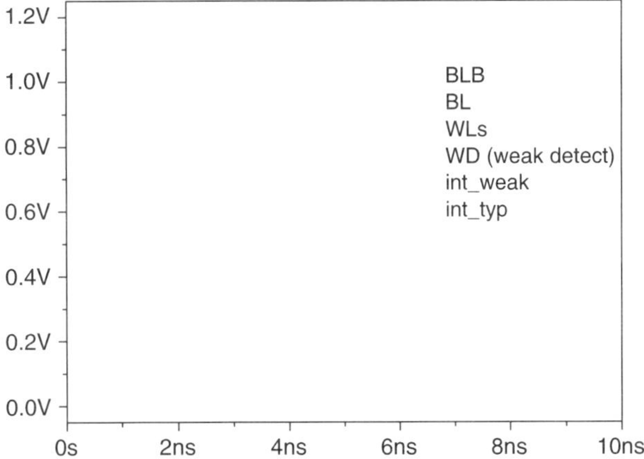

Fig. 5.26 Voltage dynamic of node B and other signals for weak cell (int weak) and a reference cell (int typ) for a 0/1 ratio of 5/3

all *n* word lines are asserted followed by the enabling of the weak detect (WD) pulse shortly after that to enter the weak detection mode. The cell's state can be inspected at around 6 ns point on the time scale.

When the ratio of 0/1 states is 4/4, the bit line-bar voltage drops to around 0.6 V ($V_{DD}/2$) but the weak cell's state does not flip. If the ratio of 0/1 states is 3/5, V_{BLB} rises up to about 1 V (Figure 5.25). This voltage strengthens the weak cell and helps it remain in its logical "1" state (bold solid line *int weak* in Figure 5.25). With a 5/3 ratio of 0/1 states (Figure 5.26) V_{BLB} drops to about 400 mV forcing the weak cell to flip states (bold solid line *int weak* in Figure 5.26).

Figure 5.27 demonstrates the detection capability of the RCRPT test technique. The resistance value of the node *A* to node *B* resistor for imitating a weak cell was varied in the range from 100 to 500 kΩ and a 0/1 ratio of 5/3 was used.

Signal *int weak* in Figure 5.27 represents node *B* of the weak cell. After applying the test sequence, the weak cell flips for resistance of 100 and 200 kΩ between node *A* and node *B*. In this case, the SNM of the weak cell is too small to resist the overwriting disturbance and the cell is overwritten. Note that the cell does not flip when the resistor values of 300, 400, and 500 kΩ because in this case the SNM is large enough to resist the test disturbance. Although it is still possible to force such a cell to flip by choosing a different 0/1 ratio of *n* cells, i.e., by digitally programming the stability detection threshold of the CUT.

The RCRPT successfully detects the defects which were simulated as resistive opens in one or both of the load PMOS transistors of the CUT. Hence, the RCRPT is confirmed to be effective in detection of both the stability and the Data Retention Faults in SRAM cells.

The second possible hardware implementation of the RCRPT test technique is shown in Figure 5.28. It has two cross-coupled pull-up PMOS transistors ($Q1$, $Q2$), pull up and pull down transistors ($Q3$, $Q4$) tied to the bit lines, a CMOS switch

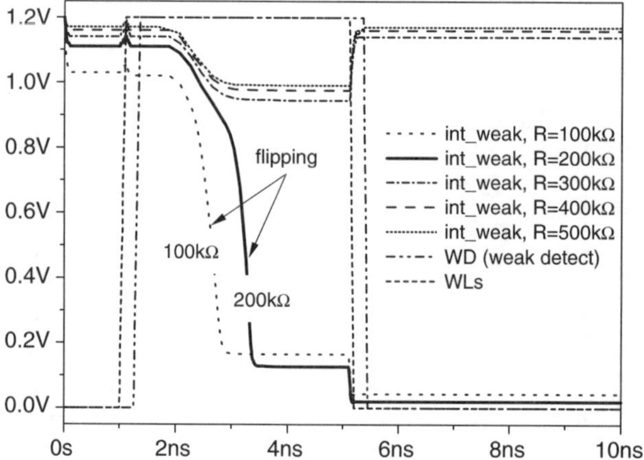

Fig. 5.27 Detection capability of the proposed DFT (implementation-1) for a 0/1 ratio of 5/3

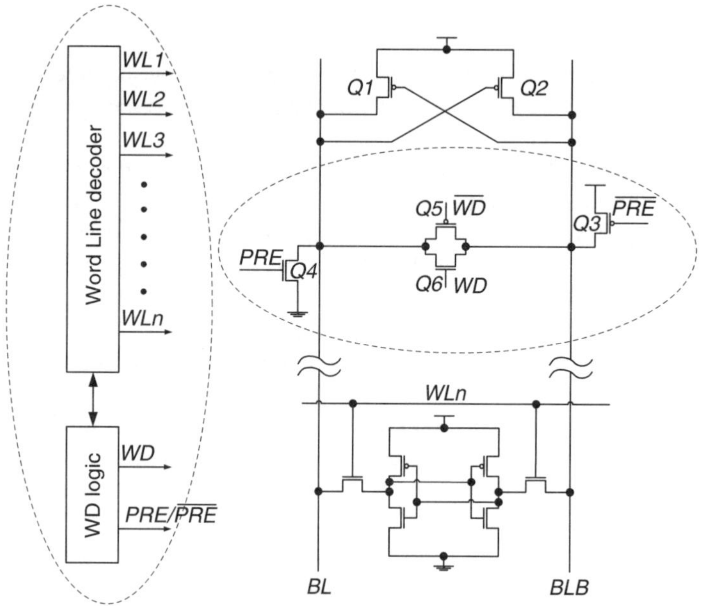

Fig. 5.28 Hardware implementation of the programmable RCRPT, ver. 2 [6]

($Q5$, $Q6$) to short the bit lines together and special logic to enable n word lines simultaneously and to test-precharge the bit lines.

The sequence of steps to carry out this implementation is similar to that for the first implementation. However, in contrast with the previous implementation, the *BL* is precharged to the ground, and the *BLB* – to V_{DD} rather than both bit lines are precharged to V_{DD}. Applying the bit line capacitances precharged in such a fashion works in a similar way to a write operation. After precharging the bit lines by enabling $PRE/\overline{PRE}$, all n word lines are enabled and the $WD/\overline{WD}$ pulse is applied. The bit line potentials redistribute, causing the weak cells to flip states. Care should be taken to ensure that $PRE/\overline{PRE}$ and WL are not overlapping. Otherwise, if such an overlap is sufficiently long, even the good cells will flip, since a long overlap will effectively turn precharge transistors into write drivers.

5.4.2.5 Programmable Read Current Ratio Technique with Floating Bit Lines

A ratio of cells storing a certain number of "0"s and "1" and connected to the same column in an SRAM array can also be applied to modify the potentials of *floating* bit lines for the purpose of creating a test disturbance. The RCRFBL SRAM cell stability test technique proposed in [116] is based on this idea.

The flow diagram shown in Figure 5.29 depicts the sequence of steps necessary to implement the RCRFBL technique. The test sequence starts with writing a 0/1 ratio to the group of n cells. Next, the bit lines are precharged and equalized as in normal operation, followed by disabling the precharge transistors and leaving the bit

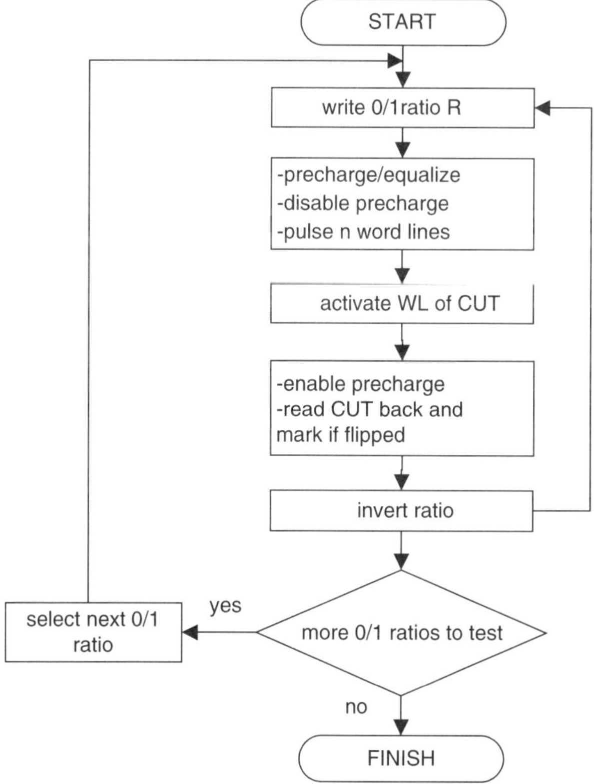

Fig. 5.29 The flow chart of the Programmable Read Current Ratio Technique with Floating Bit Lines (RCRFBL) Technique [116]

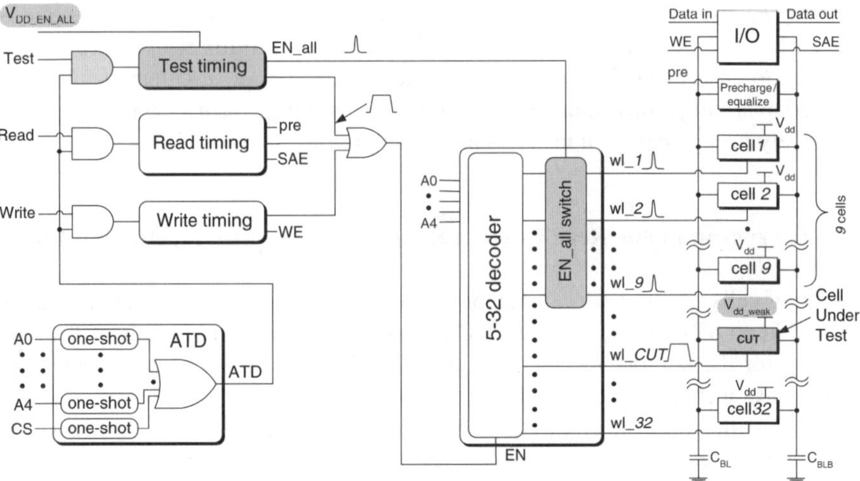

Fig. 5.30 Block-level diagram of the test chip containing asynchronous SRAM and the circuitry for the RCRFBL programmable SRAM cell stability detection technique [64, 116]

lines floating. Then, all n word lines of the cells forming the ratio R are pulsed with a short pulse. Depending on the chosen ratio R, each of the bit lines will discharge to a certain potential. In the next step, the word line of the CUT is enabled. During this step the modified bit line potentials, which are below the normal precharged values are applied to the CUT. This provides a disturbance that is aimed at testing the stability of the CUT. The test sequence is then continued with a normal read of the CUT. If the CUT has flipped, it is marked as weak. The complementary ratio R is then written into the n cells and the test sequence is repeated for the inverted ratio. The test loops until all the required ratios are tested.

One of the possible implementations of the RCRFBL technique is shown in Figure 5.30. The weak cell detection starts by determining the minimal acceptable cell stability. The cells with the SNM below that minimum must flip when connected to the partially discharged bit lines. The required bit line potentials are formed by writing n cells in the column with the necessary ratio R, precharging the bit lines and then applying a short pulse to the n word lines simultaneously enabled connecting in parallel n cells of the same column. Access transistors of each side of the n cells share a common gate and a common bit line nodes. The other terminal of each of the access transistors is connected either to the ground or to V_{DD} through the corresponding driver or load transistors of their corresponding cells. The resulting potential on each of the bit lines is a function of the chosen ratio R of the cells carrying "0"s and "1"s connected to the bit line and the pulse width of the word line pulse that enables the n cells forming this ratio.

When the n word lines are enabled, the capacitance of each bit line discharges according to the time constant created by the corresponding equivalent path. If the bit lines discharge too much, then upon the enabling of WL_{CUT} pulse, V_{TEST} can drop below the meta-stable point VM_{good}. In this case even the good cells with acceptable SNM will flip. To prevent the situation when V_{TEST} is equal or below the meta-stable point VM_{good}, the pulse width of the pulse enabling n word lines should be shortened or the ratio R should be reduced. For $R = 0.5$, the bit line voltage will be approximately equal. If $R \neq 0.5$, the corresponding path resistances to V_{DD} and the ground will be different and thus the bit lines discharged to different levels. After the bit lines have been preconditioned, the word line of the CUT is activated. Since the bit lines are left floating, effectively a read operation is performed on the CUT but with the bit lines precharged below the standard value of V_{DD}. If the CUT is weak, i.e. has inadequate SNM, then reading it with a reduced bit line voltages will cause it to flip. By controlling the degree of the bit line discharge one can shift the pass/fail threshold of the test. Both the ratio R and the pulse width $wl_1 - wl_n$ pulse can be digitally reprogrammed to set a new weak cell detection threshold.

In practice, one is free to use various arrangements to form ratio R. Ratio R can be formed either by the regular cells from the same column, or by external dedicated cells, or by a combination of the above. To improve the resolution of the proposed detection technique, the number of cells n forming the ratio R can be increased. Larger n will also help to mitigate the effect of the possible read current mismatch among the n cells. To further improve the reliability of the RCRBLF technique, two groups of n cells can be used in each column. In this case, either group of n cells can be used to test the cells in the column. One group of n cells can be used to ensure the stability of the cells comprising the other group of n cells.

Higher capacitance of the bit lines will help to achieve a higher detection accuracy as well as fine-tune the pass/fail resolution of the technique. Therefore, the proposed DFT technique may be more attractive for larger SRAM arrays with more bits per column and therefore, having more capacitive bit lines.

A test chip in CMOS 0.18 μm technology was designed to verify the RCRBLF technique. A microphotograph of the test chip is shown in Figure 5.31.

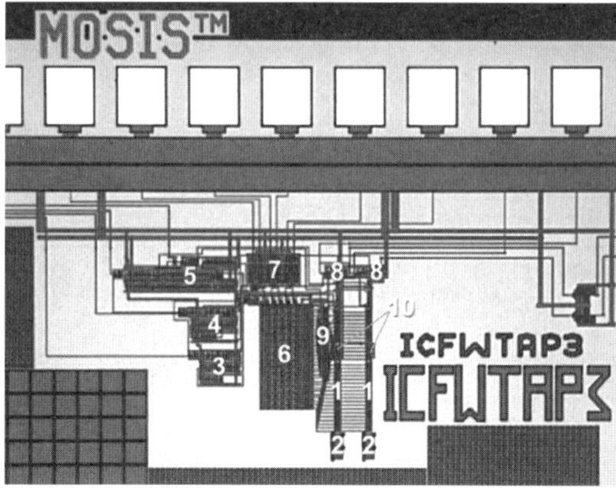

Fig. 5.31 A microphotograph of the test chip used to verify the RCRBLF programmable SRAM cell stability detection technique [64, 116]

The test chip has an asynchronous SRAM. It includes two columns of 32 cells each (1) with extra 200 fF capacitors to imitate a column with a larger number of cells (2) connected to each of the bit lines. There are self-timed blocks to provide read (3), write (4) and test (5) timing. Other blocks are the address decoder (6), address transition detector (7), sensing and writing circuitry (8), word line switches (9) and a set of weak cells (10).

Nine regular SRAM cells was formed the ratio *R* in the test chip in Figure 5.31. To enable nine word lines simultaneously, the test chip row address decoder was modified to include nine switches (9) on the first nine word lines. In practice, the switching function can be performed by two-input OR gates between the post-decoder and the word line buffers. When activated by a pulse coming from the test timing block, all nine word lines are pulled up simultaneously. The pulse can be formed locally by a simple one-shot circuit. The width of the pulse activating all nine word lines of the *n* cells in the ratio *R* must be short enough, so that the word line is deactivated before any of the *n* cells forming ratio *R* have flipped.

Figure 5.32 shows that depending on the number of cells with "0"s among the nine cells comprising ratio *R*, the same pulse width of the pulse enabling all nine word lines will discharge the corresponding bit line to a greater or a lesser extent. To control the pulse width of this pulse in the test chip, an external voltage $V_{DD\_EN\_ALL}$ was used. It supplies the delay chain in a one-shot circuit and thus modulates the width of the produced pulse. Bit line potentials as a function of $wl\_1 - wl\_n$ pulse width are shown in Figure 5.33. The required pulse width can also be specified and fixed by proper sizing of the delay chain inverters. In this case, adjusting of the detection threshold is done only by changing ratio *R*.

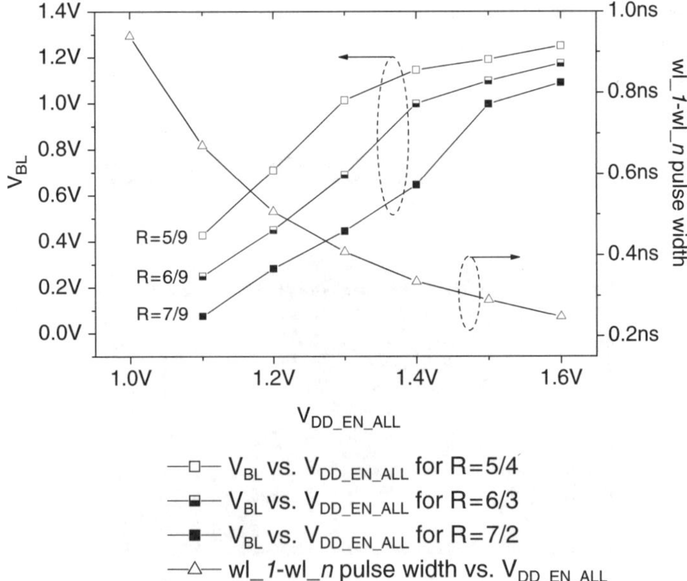

Fig. 5.32 Bit line voltage and the pulse width of $wl\_1 - wl\_n$ pulse as a function of $V_{DD\_EN\_ALL}$ (post-layout simulation results)

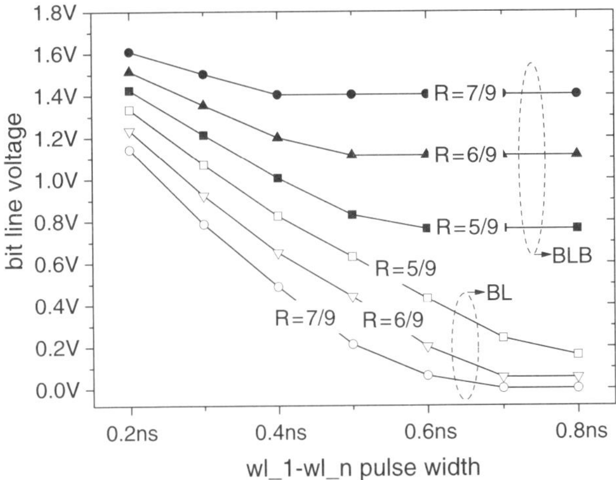

Fig. 5.33 Bit line voltage as a function of $wl\_1 - wl\_n$ the pulse width (post-layout simulation results)

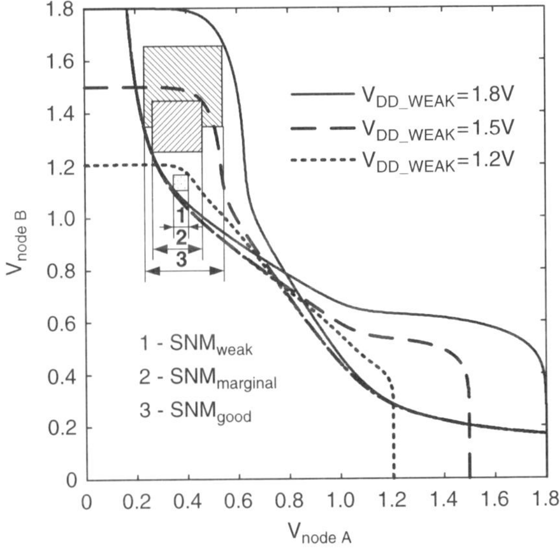

Fig. 5.34 Dependence of the VTC shape of an SRAM cell on the cell supply voltage (V_{DD_WEAK})

After the bit lines are discharged to a certain extent, defined by the chosen ratio R and the pulse width of n word lines, the word line of the cell under test is activated and the reduced voltages of the floating bit lines are applied. From Figure 5.34 one can see that a good (defect-free) cell has significantly larger SNM than a weak (defective) cell. Therefore, a weak cell will flip when read-accessed with a lower V_{BL} applied to the node storing a "1" and be detected, whereas a good cell will withstand this stress.

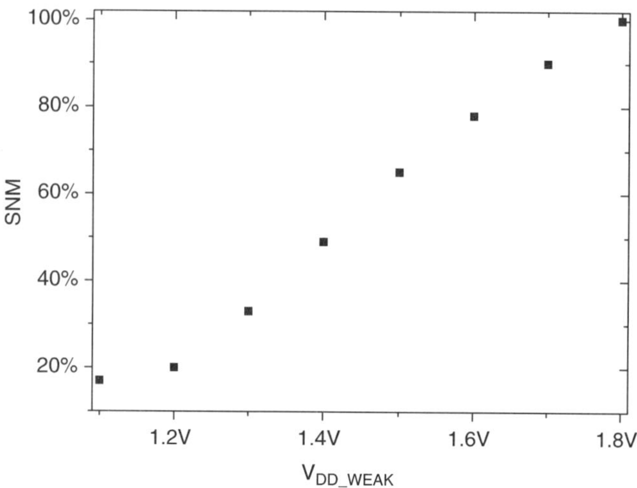

Fig. 5.35 SNM as a function of the cell supply voltage $V_{DD\ WEAK}$ (post-layout simulation results)

To imitate weak cells in this work, several cells with a separate supply voltage $V_{DD\ WEAK}$ (Figure 5.31(10)) were used. These cells had an independent power supply. Figure 5.35 shows that reducing V_{DD} of an SRAM cell reduces the SNM of the cell and thus - the cell stability. For instance, to simulate a CUT with 50% of SNM, one needs to reduce the power supply voltage of that CUT to approximately 1.4 V.

Test stimuli were provided by an Agilent 93000 SoC series tester. For each combination of $V_{DD\ WEAK}$ and $V_{DD\ EN\ ALL}$, a predetermined ratio R of "0"s and "1"s was written into the n cells. $V_{DD\ WEAK}$ and $V_{DD\ EN\ ALL}$ were swept from 0.9 V to 1.8 V. After applying the test sequence shown in Figure 5.29 the state of the CUT was registered. Figure 5.36 presents the Shmoo plots for ratio R of five "0"s and four "1"s (a), six "0"s and three "1"s (b), and for seven "0"s and two "1"s (c) among the nine cells. The black rectangles represent the combinations of $V_{DD\ WEAK}$ and $V_{DD\ EN\ ALL}$ at which the CUT flipped and the white rectangles present the combinations where the CUT maintained its data.

As it is apparent from analyzing the Shmoo plots, the detected degree of cell weakness for every fixed value of $V_{DD\ EN\ ALL}$ is different and depends on the set ratio R. For example, if $V_{DD\ EN\ ALL}$ is fixed at 1.2 V, then for $R = 5/9$, $6/9$ and $7/9$, the CUT will flip its state after the application of the proposed test sequence at $V_{DD\ WEAK} = 1.06$ V, 1.24 V and 1.36 V respectively (white lines and circled dots in Figure 5.36(a)–(c)).

The measurement results for this case are summarized in Table 5.2 and in Figure 5.36(d). It shows that by programming ratio R to be $5/9$, $6/9$ and $7/9$ and applying the RCRBLF test sequence, the weak cells detection threshold is programmed from 18% to 46% of the nominal SNM.

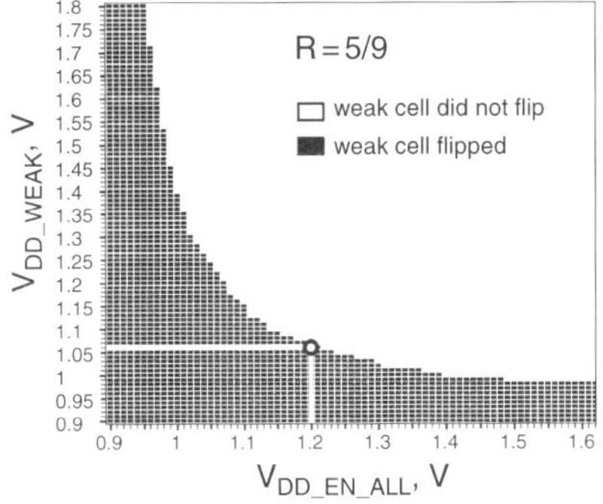

(a) Shmoo plot for $R = 5/9$; a circled dot represents the detection threshold for $V_{DD\_EN\_ALL} = 1.2V$

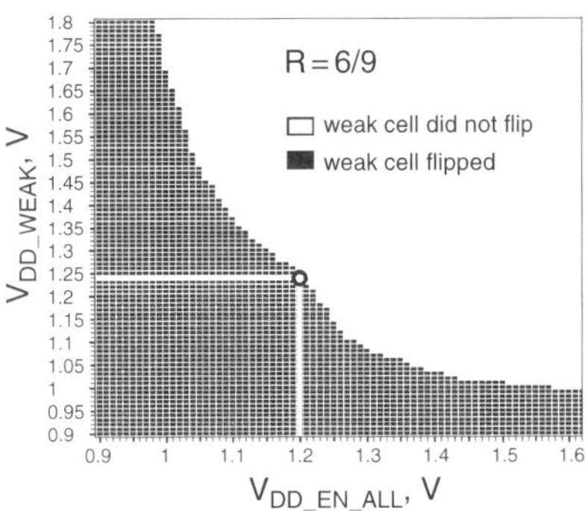

(b) Shmoo plot for $R = 6/9$; a circled dot represents the detection threshold for $V_{DD\_EN\_ALL} = 1.2V$

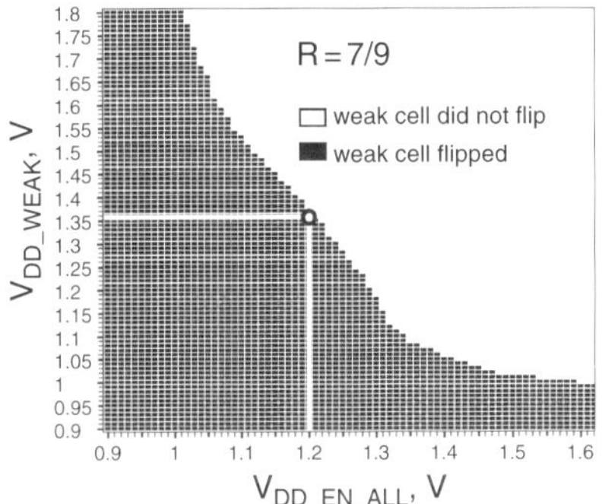

(c) Shmoo plot for $R = 7/9$; a circled dot represents the detection threshold for $V_{DD\_EN\_ALL} = 1.2V$

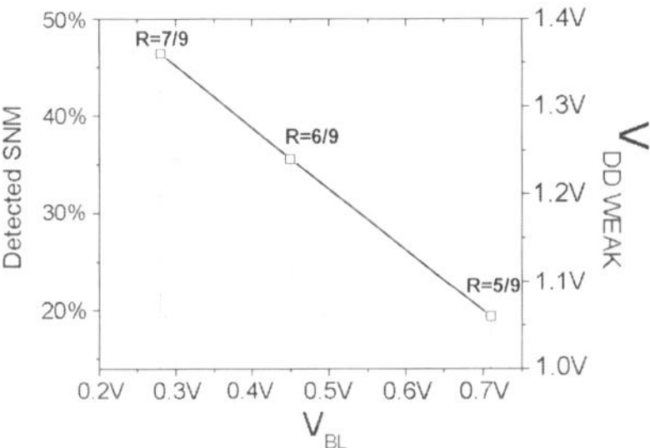

(d) Detection capability of the proposed DFT technique for $V_{DD\_EN\_ALL} = 1.2V$. Programming R from 5/9 to 7/9 changes the weak cells detection threshold from 18% to 46% of the nominal SNM

Fig. 5.36 Shmoo plots for ratios $R = 5/9$, $R = 6/9$ and $R = 7/9$ (Figure 5.36(a)–(c) respectively) and a summary for $V_{DD\_EN\_ALL}$ fixed at 1.2 V (Figure 5.36(d)) that corresponds to 500 ps pulse width of $wl_1 - wl_n$ pulse (Figure 5.32)

Table 5.2 Detection capabilities of the RCRBLF stability test technique

Ratio R	R = 5/9	R = 6/9	R = 7/9
V_{BL} (at $V_{DD\_EN\_ALL}$ = 1.2 V)	0.28 V	0.45 V	0.70 V
Detected V_{DD_WEAK}	1.06 V	1.24 V	1.36 V
Detected SNM (% of nominal SNM)	18%	36%	46%

5.4.2.6 Programmable Word Line Pulsing Technique

Yet another approach to creating the test stress for stability test was proposed by Pavlov et al. [117]. Conceptually, this technique uses the test stress, which is created by reading the CUT with one of the bit lincs being partially discharged. To achieve a reduced potential on one of the bit lines, one of the embodiments of this technique uses the read current pulses of a reference cell. Once one of the bit lines is discharged to a desired programmed potential, the CUT is read. The asymmetrical precharge levels on the bit lines create a stress on the CUT. If the SNM of the CUT is below the test threshold, such a CUT will flip its state and can be detected by a subsequent read operation. This test method is known as the Programmable Word Line Pulsing Technique (PWLPT). Below, we will explain the PWLPT in detail.

The principle behind the PWLPT is clear from the circuit shown in Figure 5.37. It represents a section of a column in an SRAM array with two identical cells. Let us assume that the top cell is the reference cell and the bottom cell is the Cell Under Test (CUT).

Suppose node A of the CUT is storing a "0", while node B is storing a "1" and resistors $R1$ or $R3$ represent opens in the pull-up path of the cell. It is well known that the highly-resistive opens in the pull-up path of an SRAM cell can cause Data Retention Faults if the pull-up current through $R1 - Q4 - R3$ in Figure 5.37 fails to compensate for the off-state leakage of the driver transistor $Q2$. Given sufficient time, the leakage current of $Q2$ will discharge the capacitance of node B. Once V_{node_B} has crossed the switching threshold of the cell, the cell will flip, which can be detected by a subsequent read operation. This situation is similar to a very slow weak overwriting of the cell under test. For higher values of the $\frac{I_{pull-up\_path}}{I_{leakage\_driver}}$ ratio,

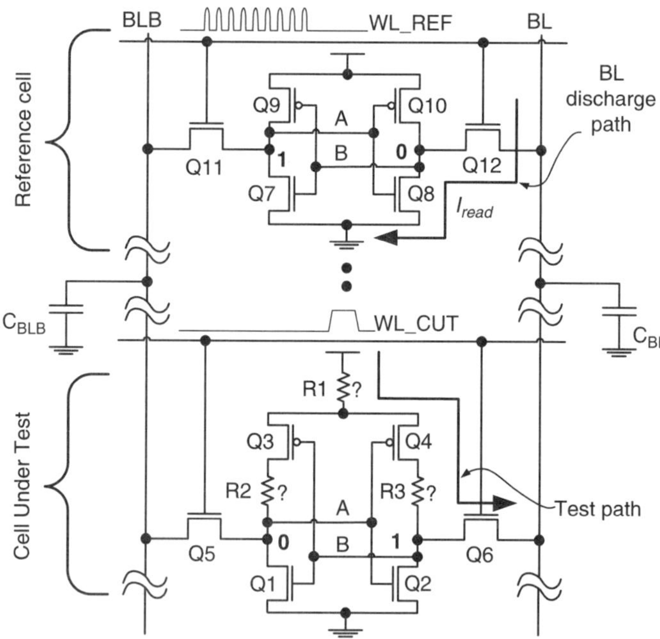

Fig. 5.37 The concept of the Programmable Word Line Pulsing Technique (PWLPT). R1 represents a symmetric defect; R2, R3 - asymmetric defects

the time required to discharge node B and flip the CUT can be longer than is economical for the delay part of the Data Retention Test (DRT) to detect such a defect. If the pull-up current is even marginally greater than the off-state leakage current of $Q2$, the cell may escape the DRT even with extended test delay periods and elevated temperature. Memories with such highly unstable cells may be shipped to the customer and cause customer returns.

The gate voltage of $V_{G\_Q1} = V_{node\_B}$ will not change significantly unless I_{leak_Q2} $(R1 + R3)$ is significant, i.e. at least one of the weak opens $R1$ or $R3$ qualifies for a highly-resistive open. So, node A and the gate of $Q4$ will remain at "0" and the effective pull-up current for node B of the CUT will be proportional to the equivalent resistance of the path $(R1 - Q4 - R3)$.

Suppose we can freely change the potential on the bit line (BL) and set it to be $V_{DD}/2$ (in our example - 0.6 V) while the complementary bit line (BLB) remains fully precharged to the supply voltage 1.2 V. After enabling WL_CUT, $Q6$ will pass the reduced V_{BL} onto node B. Node B potential is proportional to the ratio $\frac{I_{Q4}}{I_{Q6}}$. The overwrite condition for node B is ensured if we can pull node B below the switching threshold of the inverter formed by transistors $Q1$ and $Q3$. Since the effective pull-up drive of node B is weakened by the defect resistance $(R1 + R3)$ in its pull-up path, the overwrite condition is met earlier and the weak cell is overwritten, whereas a good cell with $(R1 + R3) \to 0$ in the same situation would withstand the same stress.

The stress level setting in the PWLPT technique is based on the realization that a precharged bit line BL coupled through the access transistor $Q12$ to the node B of the reference cell carrying a "0" (Figure 5.37) is gradually discharged by the I_{read} of the reference cell. The discharge rate can be expressed by Equation (5.1) and is a function of the *cumulative total duration* that the word line of the reference cell has been enabled.

$$\Delta V_{BL} = \frac{I_{read} \times t_{WL\_REF\_pw}}{C_{BL}} \tag{5.1}$$

where: ΔV_{BL} - discharge of the bit line after each enabling of the WL_REF; I_{read} - cell read current of the reference cell; $t_{WL\_REF\_pw}$ - pulse width of the reference cell word line pulse; C_{BL} - bit line capacitance.

Figure 5.38 shows the waveforms of the bit line voltage discharge after each of the 32 WL_REF pulses. Since ΔV_{BL} is inversely proportional to C_{BL}, i.e. the bit line discharge rate is slower and the precision of the bit line voltage setting is higher for the higher values of the bit line capacitance. One of the ways to increase the bit line capacitance and hence the accuracy of the test stress in the SRAM architectures with local and global bit lines is by enabling all the bit line MUX transistors that connect the local bit lines to the global in a given column.

The flow diagram of the proposed weak cell detection technique is shown in Figure 5.39. The test procedure writes the opposite data backgrounds into the reference cell and the CUT. After the normal precharge is completed, the word line of the reference cell is enabled N times. Each enabling of the WL_REF gradually dischares the bit line (Figure 5.38). Then, the word line of the CUT is enabled to test the stability of the CUT. Next, the CUT is read normally and its state is inspected. To ensure that asymmetric faults are covered as well, the data backgrounds

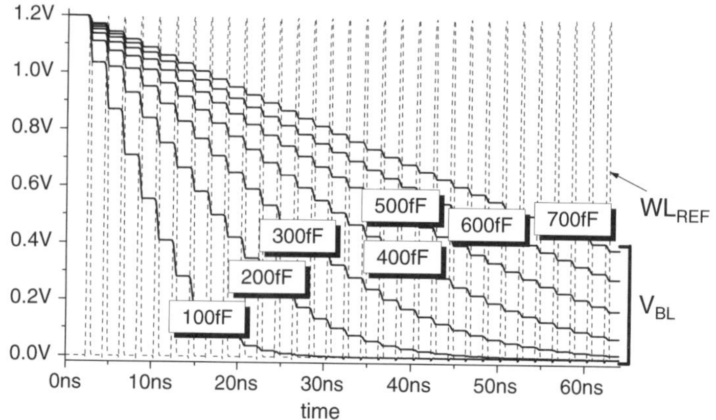

Fig. 5.38 Word line pulses of the reference cell discharge the bit line for various values of the bit line capacitance (simulation results in CMOS 0.13 μm technology with the pulse width of the reference cell word line of 410 ps

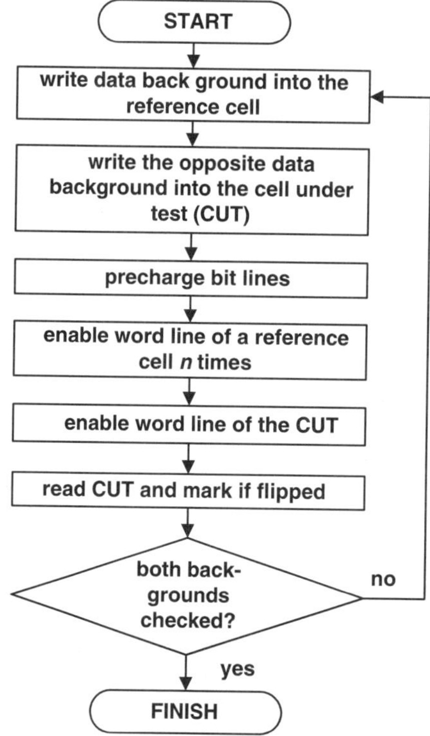

Fig. 5.39 Flow diagram of the Programmable Word Line Pulsing (PWLPT) technique for stability fault detection in SRAM cells

in the reference cell and the CUT are then inverted and the PWLPT test sequence is repeated. To further reduce test time, this test can be conducted in parallel on a word line per word line basis with one reference cell and one CUT per column.

Since the rate of the bit line discharge using a reference cell is a factor of the total duration that WL_REF has been enabled, the required degree of the bit line discharge can be achieved by several methods.

The first method is illustrated in Figure 5.38. The cumulative total duration of WL_REF can be composed of N pulses. The desired degree of the bit line discharge can be provided by changing the number of the WL_REF pulses. The second possible approach is to fix the number of WL_REF pulses N while changing their pulse width. The third approach is to hold WL_REF for a predefined time period ($N = 1$). The duration of this time period will define the discharge level of the bit line. The reference cell can be interchanged with the CUT or another cell sharing the same bit lines with the CUT. The inter- and intra-cell read current spread of the reference cells caused by the process variations can be alleviated by using a larger dedicated reference cell. Another degree of freedom can be provided by using several dedicated reference cells with different read currents. In other words, the PWLPT offers extended flexibility in setting the weak overwrite stress. To further improve the accuracy of the PWLPT, the reference cell(s) can be replaced by dedicated NMOS long-L transistors. However, this implementation entails a small area overhead.

To prove the concept of the PWLPT, we designed a synchronous self-timed 8 Kb SRAM test chip in CMOS 0.13 μm technology. The top-level layout and the block-level diagram of the test chip are presented in Figures 5.42 and 5.43, respectively.

The SRAM array has 32 columns and 256 rows in a 2 words by 16 bits architecture. The test chip features a fully self-timed control block. The read access time and power is minimized by using the replica (dummy) loop based timing technique. The Control block contains a reset-dominated Finite State Machine (FSM). When it is set by the rising edge of the clock (CLK), it disables precharge and enables the Dummy Word Line (DWL) and a regular word line. Once the single-ended Dummy Sense Amplifier (DSA) flips, it resets the FSM stopping further discharge of the bit line. Next, the Sense Amplifier Enable (SAE) is issued to latch the V_{diff} on the bit lines. The YMUX is disabled shortly after to prevent the bit lines from being discharged by the SA. More details on the principle of the replica-loop based timing is presented in Section 2.9.2.

The Chip Select (CS) input is latched on the rising clock edge ensuring that the subsequent read or write operation is applied to this chip. A high level on Write Enable Bar (WEB) enables activation of the SA and YMUX and disables the write drivers providing for a successful read operation. Conversely, a low level of the WEB switches SRAM into the write mode by blocking the SAE and YMUX and enables the write drivers. Address bit A8 (YMUX) provides an address for selection of the column to be connected to the SA by YMUX. Combined with the row address bits A0–A7, that makes for nine address bits overall.

Signals WL_REF_start and WL_CUT_start can be connected to the system CLK to initiate each of the WL_REF or WL_CUT pulses for the reference cell or the CUT, respectively. For added flexibility, the WL_REF or WL_CUT pulse width is adjustable via WL_REF_pw or WL_CUT_pw levels. These levels control the PMOS bias in the current-starved inverters used in WL_REF or WL_CUT one-shot circuits.

The formed WL_REF or WL_CUT pulses gate the accessed word line. An example of a REF cell and the CUT is shown in Figure 5.43.

The PWLPT was verified to be effective for various values of the bit line capacitance, number and the pulse width of WL_REF pulses. It can detect a wide range of resistive defects in the pull-up path of the CUT. To demonstrate the detection capabilities of the proposed technique we used twelve WL_REF pulses with the pulse width of 410 ps and $C_{BL} = 400$ fF.

It was found that if the overwrite stress is too strong, i.e. the bit line can be discharged below a certain point, which in our simulations was 0.55 V, when even the defect-free SRAM cells will flip. The area where the defect-free cells had flipped is represented in Figure 5.40 by the patterned rectangle. As was mentioned before, the rate of the bit line discharge is a function of the bit line capacitance. Depending on the bit line capacitance, a different number of WL_REF pulses is required to reach the desired degree of the bit line discharge.

Figure 5.41 shows the simulation waveforms illustrating the detection capability of the PWLPT when applied to a symmetric defect (resistive contact "1" in Figure 4.3 and $R1$ in Figure 5.43). After WL_REF was enabled 12 times, the bit line was discharged from 1.2 to 0.65 V. Figure 5.41(a) demonstrates that for $R1 = 80\,\text{k}\Omega$ the potential levels of nodes A and B of the cell have not reached the metastable point of the CUT. Therefore, a defect with 80 kΩ resistance was not detected. However, once the PWLPT encounters a defect with the resistance of 120 kΩ (Figure 5.41(b)), node B is driven low enough to cross the switching threshold of the inverter formed by transistors $Q1$ and $Q3$ of the CUT and the CUT had flipped its state. Consequently, when the CUT is read back after the application of the PWLPT test sequence, the CUT with $R1 = 120\,\text{k}\Omega$ provide incorrect data and will be marked as defective.

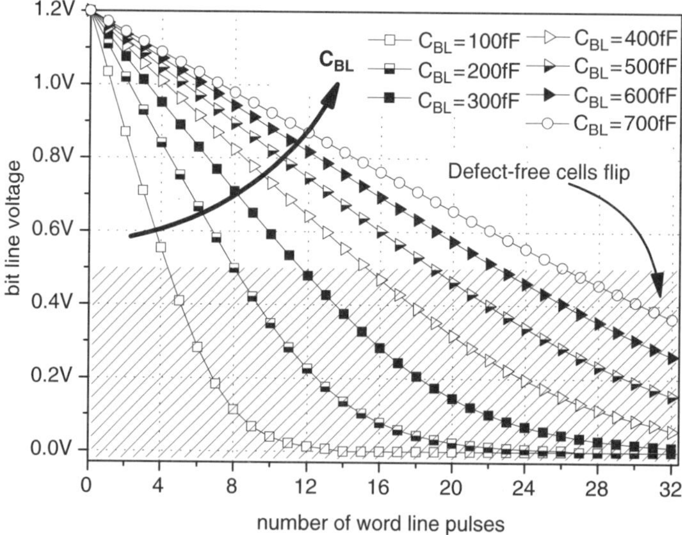

Fig. 5.40 Discharge of the bit line as a function of the number of word line pulses of the reference cell and the bit line capacitance

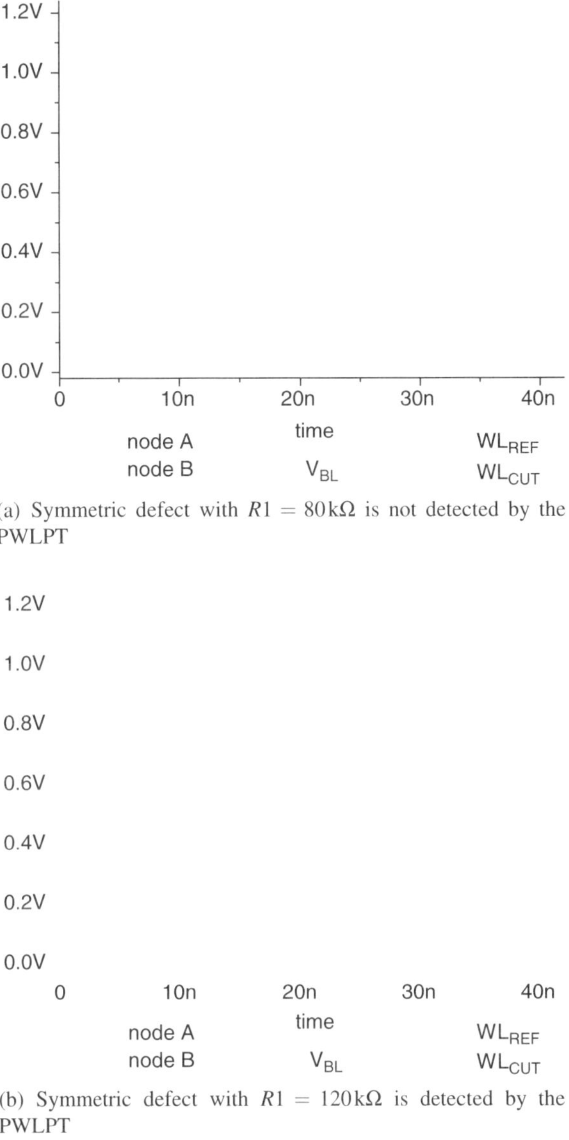

(a) Symmetric defect with $R1 = 80\,\text{k}\Omega$ is not detected by the PWLPT

(b) Symmetric defect with $R1 = 120\,\text{k}\Omega$ is detected by the PWLPT

Fig. 5.41 Example of a symmetric defect with (a) resistance of $80\,\text{k}\Omega$ (not detected) and (b) resistance of $120\,\text{k}\Omega$ (detected), using the PWLPT (CMOS 0.13 µm, $C_{BL} = 400\,fF$)

Note that the slope of the cell's VTC in the metastability region $\frac{dV_{out}}{dV_{in}}$ that is proportional to the *ac* gain of the cell is a function of an SRAM cell's SNM. The $\frac{dV_{out}}{dV_{in}}$ is steeper for a cell with a higher SNM. The resolving capability (gain-bandwidth product) of the cell in the metastable region is proportional to its SNM [20]. Therefore, if the SNM of the cell is higher, the cell will have a higher immunity against metastability and recover from the test disturbance quicker.

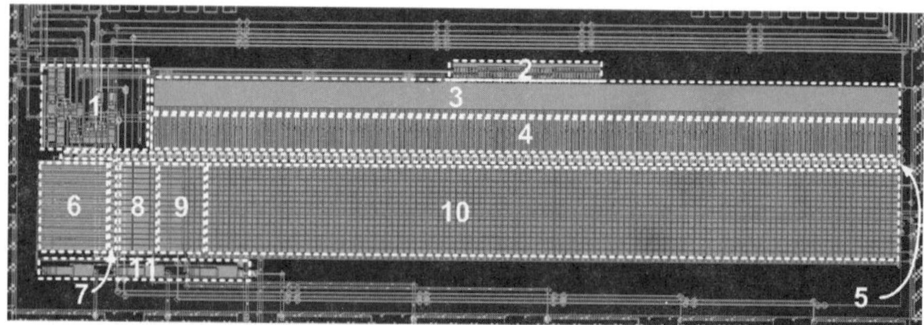

Fig. 5.42 8 Kb synchronous SRAM test chip with WLPT (CMOS 0.13 μm technology): (1) control block, (2) decoder, (3) post-decoder, (4) word line drivers, (5) dummy column and dummy SA, (6) SAs, column MUXs, write drivers and precharge/equalization, (7) dummy row, (8) reference SRAM cells for PWLPT, (9) weak SRAM CUTs for WLPT, (10) regular SRAM cells and (11) pad drivers [64]

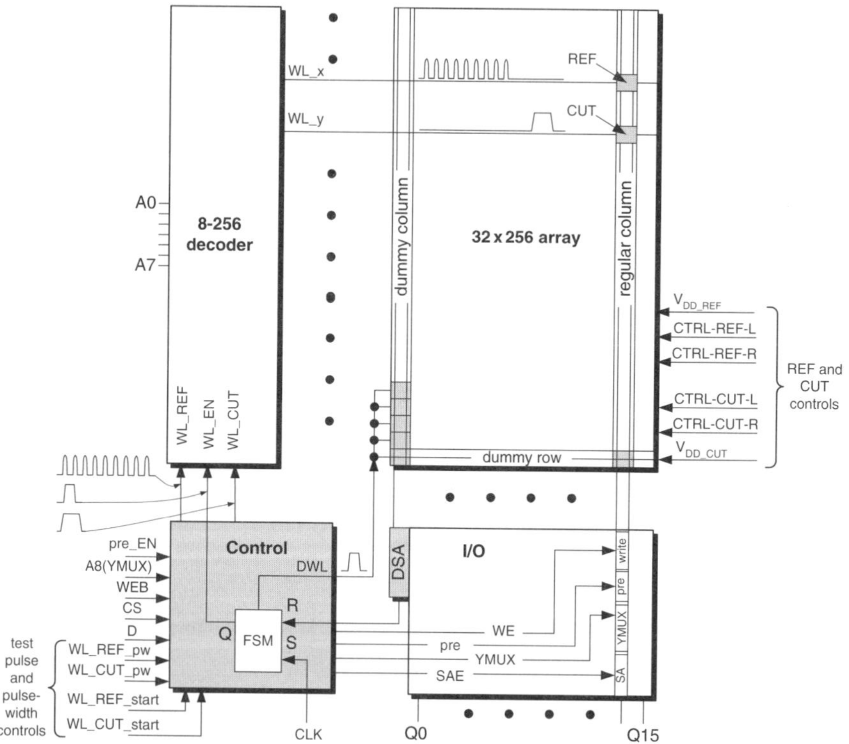

Fig. 5.43 Block-level diagram of the PWLPT test chip [64]

However, if the stability of the cell is weakened by a defect or a mismatch and its SNM is reduced, then such a cell will remain in the metastabile state longer. Thus, to extend the range of the detected defect resistance, the duration of the WL_CUT pulse should be sufficiently long to allow for the extended metastability window of the weaker SRAM cells before a stable state is resumed.

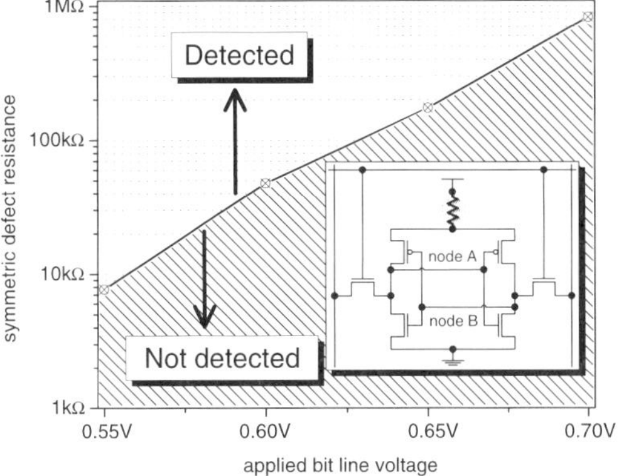

(a) Detection of a symmetric defect by the PWLPT

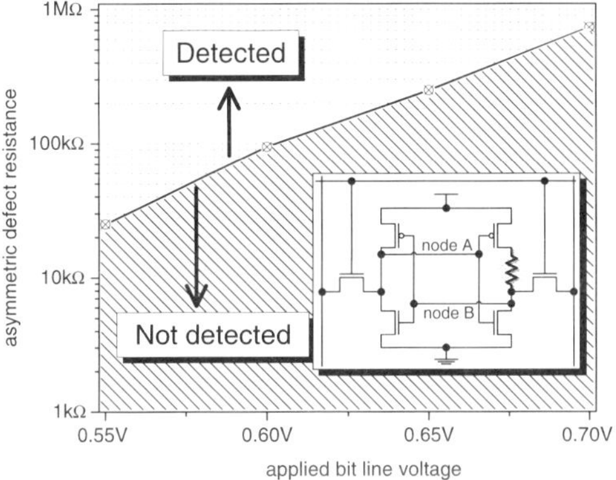

(b) Detection of an asymmetric defect by the PWLPT

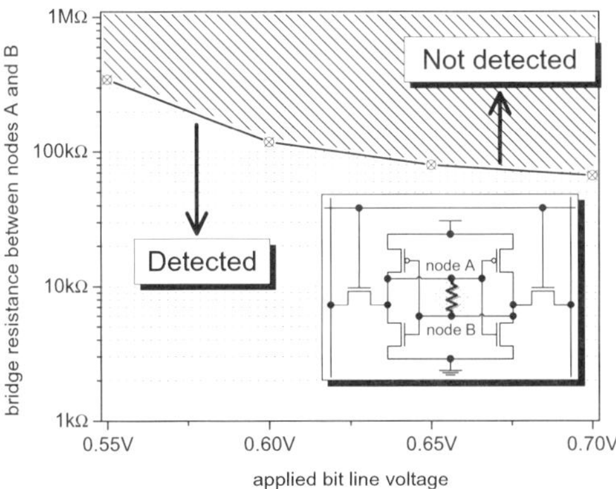

(c) Detection of a resistive bridge between the internal nodes A and B of an SRAM cell by the PWLPT

Fig. 5.44 Detection of a symmetric (a), an asymmetric (b) defects in PMOS transistors of 6T SRAM cell and a resistive bridge between node A and node B using the PWLPT (CMOS 0.13 μm, $C_{BL} = 400\,\text{fF}$)

For instance, if $R1$ is between 80 and 120 kΩ, e.g. 95 kΩ, the potentials of node A and node B will move closer to each other and to the metastable point of the SRAM cell. Our simulations showed that due to the reduced *ac* gain in this region the WL_REF pulse should be asserted for several nanoseconds for such a cell to reach a stable state.

Waveforms that are very similar to those presented in Figure 5.41 were obtained for an asymmetric defect $R3$ with $R3_{detected} = 180\,\mathrm{k\Omega}$ and $R3_{undetected} = 190\,\mathrm{k\Omega}$. The PWLPT was also verified to successfully detect highly-resistive opens in the gates of PMOS transistors.

We characterized the detection range of the PWLPT with the bit line capacitance C_{BL} set to 400 fF. Figure 5.44 summarizes our findings. The PWLPT was characterized to efficiently detect both the symmetric (Figure 5.44(a)) and asymmetric (Figure 5.44(b)) defects that cause data retention and stability faults. The PWLPT demonstrated high selectivity of the detected defect resistances. The solid line in Figure 5.44 represents the detection boundary of the PWLPT. The area above represents the detected values of the defect resistance and the patterned area represents range of the defect resistances, which is not detected.

Figure 5.44(c) demonstrates the detection of the bridge resistance between node A and node B as per the fault model representing an unstable SRAM cell, as presented in [6]. According to our estimates, a resistive bridge between the internal nodes of an SRAM cell (node A and B) is one of the more probable bridge defects in the cell. Since it also represents a symmetric negative feedback branch for the two inverters of an SRAM cell, its resistance is directly reducing the *ac* gain of the cell in metastable region, changes the shape of the VTCs and reduces the SNM of an SRAM cell. We believe that it can mimic the behavior of a multitude of resistive bridge defects in an SRAM cell. Since the SNM dependence on the bridge resistance is the inverse of that of the resistive opens, the solid line in Figure 5.44(c) exhibits inverse slope as compared to Figures 5.44(a) and (b). The higher defect bridge resistance values correspond to a more stable cell and thus the detected resistance region in Figure 5.44(c) is above the solid boundary line.

5.5 Summary

This chapter presented a comprehensive analysis of special test techniques for detection of stability faults in SRAM cells. We introduce a classification of the existing stability test techniques that helps to group them based on whether the test technique is functional (passive) or structural (active) and whether an active test technique allows for programmability of the produced test stress.

We discussed the evolution of SRAM stability test techniques from passive functional tests such as the data retention test and the low-voltage test to the active test techniques with a single and programmable stress settings. The evolution of the stability test is largely due to the continuing scaling of the semiconductor process technology. The scaled transistors and interconnects have inherently larger parameter variations and defect count. Each new process technology challenges circuit designers to create more and more flexible DFT techniques that can of separate the SRAM cells weakened by large process variations and subtle defects from healthy cells with acceptable stability.

The drawback of the single test stress level techniques is that they may under-stress or over-stress the tested SRAM cells while attempting the detection of the targeted defects. The inaccurate setting of the test stress level may result from the inability to track the process modifications. Achieving the desired trade-off between the quality and the test yield may require multiple post-silicon design iterations that target the changing process conditions following the initial design.

Thanks to the introduction of cell stability test techniques with programmable test stress, the amount of test stress can be determined/adjusted after the design stage is complete. In contrast to SRAM cell stability test techniques with a single test stress that attempt to apply the preselected test stress across wide ranges of process variations and cell defect mechanisms, programmable test techniques offer lower design effort and risk. Allowing the test stress to be controlled during the production process helps to mitigate the risk that if the fabricated SRAM cells are not operating exactly as it was intended during the design stage, the accurate stability testing still can be performed after the adjustment of the test stress. Such a risk may be a result of an incomplete or inaccurate process models during the design stage, while the manufactured SRAM array could exhibit a different or unexpected behavior. The programmability of the test stress enables balancing between the number of test escapees due to insufficient test stress and the test yield loss caused by the excessively stringent stability testing that fails some of the good SRAM cells. Choosing the correct stress for SRAM cell stability testing is a non-trivial task. Making the complex decision of drawing a line between good stable cells and defective unstable cells is often possible only after analyzing the test characterization results of a significant population of SRAM cells in high-volume manufacturing.

The two classes of the stability test techniques are illustrated by a number of state-of-the-art practical example implementations. We included detailed explanations of their circuit operation, test algorithms and defect detection capabilities. While not all of the existing SRAM stability test techniques are covered in this chapter, the discussed examples can provide a design or test engineer with a good overview of the current developments in this area and prime the reader in the field of SRAM cell stability test.

Chapter 6
Soft Errors in SRAMs: Sources, Mechanisms and Mitigation Techniques

6.1 Introduction

Technology scaling dramatically increases the sensitivity of the semiconductor devices to radiation [118]. Due to large number of cells with minimized dimensions, SRAM arrays often are the densest circuitry on a chip. The large bit count contributes to the probability that an ionizing particle will hit a sensitive node in the array and corrupt the stored data. The minimum layout dimensions reduce the storage node capacitance and thus, the critical charge Q_{crit} that can be injected by radiation and upset the SRAM cell. The shrinking supply voltages reduce the Q_{crit} even further. These factors contribute to the radiation-induced data errors that complicate building reliable SRAM arrays in nano-scaled technologies.

Radiation can create localized ionization events in the semiconductor devices either directly or as secondary reaction products. Many of these radiation-induced events create enough electron-hole pairs to upset the storage nodes of SRAM cells. Such an upset is called a *"soft" error*. While such an upset can cause a data error, the device structures are not permanently damaged. If the voltage disturbance on a storage node of an SRAM cell is smaller than the noise margin of that node, the cell will continue to operate properly maintaining its data integrity. However, if the *noise margin of a cell is not sufficient* to withstand the disturbance caused by ionizing radiation, a "soft" error will result.

Soft errors originally were discovered to be a problem for DRAMs with planar storage capacitors in the late 1970s [119]. Such capacitors stored their charge in large area two-dimensional *p-n* junctions. Due to the high collection efficiency of the radiation-induced charge, the large planar reverse-biased junctions made early DRAM cells highly susceptible to soft errors. Technology scaling and the push to improve the high soft error rate (SER) and poor pause/refresh time ratios of DRAMs necessitated the development of more compact three-dimensional storage capacitors. Compared to the old 2D planar capacitors, the new 3D capacitors had significantly smaller junction collection efficiency [120] due to the reduced volume of the *p/n* junction [121]. Despite the reduction of the Q_{crit} of a DRAM cell due to V_{DD}

scaling, it was more than compensated by the aggressive junction volume scaling of the storage capacitor. As a result, the SER of a DRAM *bit cell* is reducing by about 4x per technology generation [121]. This DRAM SER reduction is, however, being offset by the similar DRAM bit count growth rate at a *system* level. Because larger DRAM arrays are statistically more susceptible to soft errors, the resulting DRAM SER at the *system* level has remained relatively stable over many recent technology generations.

Early SRAMs were more robust to soft errors than DRAMs due to the feedback mechanism that helps to maintain the state of an SRAM cell. In an SRAM cell both the capacitance of the storage node and the restoring current provided by the pull-up or driver transistors contribute to the cell critical charge. With technology scaling, the SRAM cell area and thus, the junction area of the storage nodes are shrinking (Figure 6.1). While it could reduce the cell junction leakage, it also reduced the storage node capacitance. Simultaneously, the transition from the constant voltage scaling to the constant electric field scaling resulted in aggressive V_{DD} scaling. Both these factors directly contribute to the reduction of the resulting Q_{crit} and the increasing probability of soft errors leading to higher SER levels. With each successive technology generation, the reductions in cell collection efficiency due to shrinking cell depletion volume was compensated by the reductions of V_{DD} and storage node capacitance [121]. Therefore, soft errors have recently become a growing issue in ultra high-density large embedded SRAMs operating at low voltages. The SRAM bit SER is shown to have saturated for technology nodes beyond 0.25 μm [121] due to the saturation in V_{DD} scaling, reductions in junction collection efficiency of highly doped *p-n* junctions and the increased charge sharing between the neighboring nodes. However, the exponential growth of SRAM cell count in modern CPUs and DSP processors has led the SRAM system SER to increase with each technology generation.

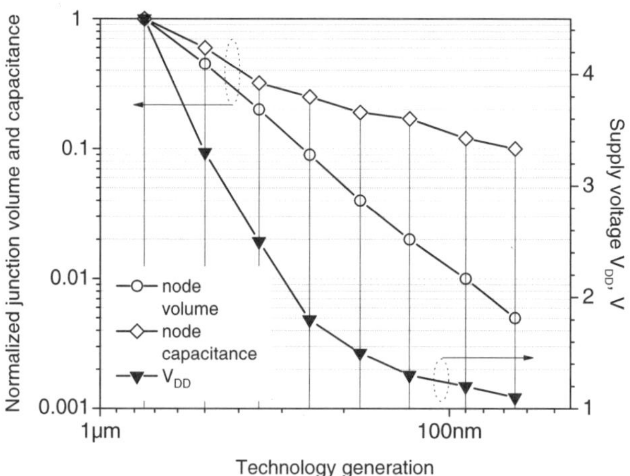

Fig. 6.1 The normalized junction volume and capacitance of SRAM storage nodes and the power supply voltage V_{DD} scaling as a function of the technology generation [121]

Combinational logic circuits may mask the propagation of a data value corrupted by a soft error because other inputs of a combinational gate may gate such a corrupted value. However, soft errors caused by energetic particles in SRAMs are always latched and thus lead to a data error. The only way to correct such data error is to rewrite the correct data into the cell. The reliability impact of the SER of SRAM arrays is estimated to be several times higher than the impacts of all other reliability factors combined. While most hard failure effects are characterized by the product failure rates of the order of 1–100 FIT, the SER of low-voltage SRAM can reach up to 1,000 FIT/Mb [122].

6.2 Soft Error Mechanism

Every ionizing particle is characterized by a specific value of the Linear Energy Transfer (LET). For instance, one electron-hole pair is generated per every 3.6 eV lost by an ion in a Si substrate [123]. LET depends on the energy and the mass of the ionizing particle as well as the properties of the penetrated material. A higher LET is imparted by heavier particles with higher energy in denser materials. The collection of the charge generated by an ionizing event occurs within a few micrometer of the *p-n* junction. The collected charge Q_{coll}, which is a function of the type of the ionizing particle, its trajectory and its energy value while it is traveling through or near a *p-n* junction, ranges from $1fC$ amounts to hundreds of fC.

The critical charge Q_{crit} required to upset a data node and flip an SRAM cell can be defined as [124, 125]:

$$Q_{crit} = \int_0^{\tau_{flip}} I_D dt = C_{node} V_{node} + I_{restore} \tau_{flip} \qquad (6.1)$$

where C_{node} and V_{node} (V_{node}=V_{DD} for a CMOS SRAM cell) are the capacitance and the voltage of the affected node respectively, $I_{restore}$ is the restore current provided by the pull-up path (PMOS transistors in case of a 6T SRAM cell), and τ_{flip} is the time required for the feedback mechanism to take over from the ion's current and flip the cell.

In a 6T SRAM cell, the data node storing a "1" is the most susceptible to single event upsets caused by ionizing particles. The Q_{crit} of a $0 \rightarrow 1$ transition where the NMOS pull-down transistor counteracts the radiation-induced current spike is reported to be ~22X larger than that to cause a $1 \rightarrow 0$ transition [126]. The state of the "1" storage node in an SRAM cell is supported by a relatively weak PMOS pull-up transistor. This results in a smaller radiation-induced current necessary to cause a $1 \rightarrow 0$ transition and upset the data node storing a "1". Therefore, the radiation immunity of an SRAM cell is defined by the Q_{crit} of the data node storing a "1".

An ionizing particle that strikes or passes in the immediate proximity of the reverse-biased junctions of node "1" can result in the junction collecting the generated electron-hole pairs. The net effect of such a disturbance is similar to overwriting the struck SRAM cell with the write current that is proportional to the collected

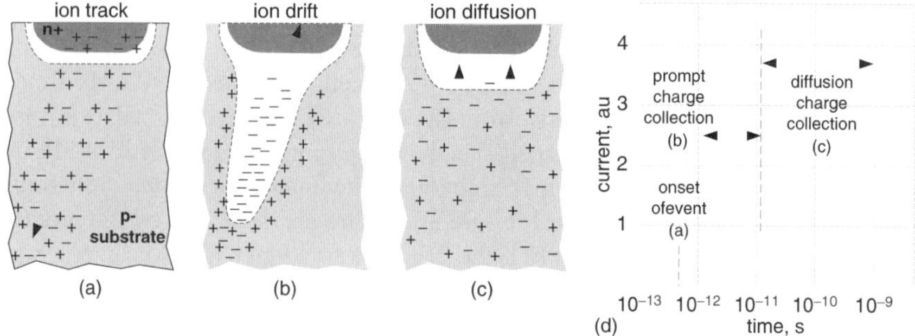

Fig. 6.2 Generation of electron-hole pairs in a reverse-biased *p-n* junction: (a) ionization of substrate atoms on the path of the striking particle; b) formation of a funnel-shaped depletion region; (c) drift charge collection is superseded by the diffusion charge collection and (d) the time line and the current pulse shape for each of the phases [121, 123]

charge. Figure 6.2 illustrates the phases of charge generation and collection in a reverse-biased *p-n* junction. At the onset of an ionizing radiation event, ionizing particles form a cylindrical track of electron-hole pairs with a sub-micron radius with high carrier concentration (Figure 6.2(a)). When the path of the ionizing radioactive particles lays through or near a depletion region of a reverse-biased *p-n* junction, the generated electron-hole pairs are rapidly separated and collected by the electric field creating a large current transient at the affected node (Figure 6.2(b)). Note that the shape of the depletion region is modified by the ionizing particles into a funnel. The funnel effectively extends the area of the depletion region and greatly increases the drift collection efficiency by extending the high electric field of the depletion region deeper into the substrate. The funnel distortion of the depletion region is inversely proportional to the substrate doping concentration. Lightly doped substrates will show a stronger distortion and thus, a higher drift charge collection compared to a more heavily-doped substrate. The current spike caused by the rapid drift charge collection phase is completed within tens of picoseconds. It is then superseded by the diffusion collection phase that can last for hundreds of nanoseconds until all the excessive charge generated by the ionizing particle has been collected, diffused and/or recombined (Figure 6.2(c)). The resulting current pulse is shown in Figure 6.2(d).

The amount of collected charge Q_{coll} and thus, the probability of the ionizing particle to create a soft error depends on the proximity of the path of an ionizing particle to the reverse-biased *p-n* junction. Another factor is the proximity of other reverse-biased junctions that can share the generated charge and decrease the current spike on a particular circuit node. The effect of an ionizing particle striking a circuit node is extremely difficult to model. Charge collection is a strong function of the struck node potential, which depends on external loading (e.g., a struck device alone vs. a struck device in a circuit). It is reported that the critical charge (Q_{crit}) required to upset an SRAM cell is often not directly correlated to the critical LET required to upset such a cell. Moreover, in some cases more charge may be collected at the data storage nodes from strikes that do not cause an upset that from strikes that do cause an upset. An upset occurs in an SRAM cell when the recovery time of the cell τ_r exceeds the feedback time τ_f. Since charge may continue to be collected either after the cell has already been upset or after it has recovered, the total collected charge Q_{coll} should not be used to find the critical charge Q_{crit} [127]. A more accurate measure can be represented as the significant charge Q_{sig} that can be expressed as:

$$Q_{sig} = \int_0^T I_{drain} dt$$

where $T = \tau_r$ if the cell recovers and $T = \tau_f$ if the cell upsets. In other words, the Q_{sig} is a portion of the total collected charge Q_{coll} which is responsible for causing an upset or an attempt on an upset (in case a struck SRAM cell recovers) [127].

6.3 Sources of Soft Errors

It has been established that SER in semiconductor devices is induced by three different types of radiation [118, 128]:

- Alpha particles
- Neutron-induced fission of ^{10}B
- High-energy cosmic rays

In the following sections we will address each of the SER radiation sources as well as possible measures that help to mitigate their potential for generating soft errors in SRAMs.

6.3.1 Alpha Particles

One radiation source that can cause soft errors in packaged SRAMs comes from alpha particles that are generated by the small traces of radio-active impurities in the package materials [118]. Alpha particles are ionized helium atoms ($^4He^2$) with 4–9 MeV energy range. They are emitted when unstable radio-active isotopes of certain chemical elements decay to the lower energy states. The dominant sources of α-particles in materials that are characterized by the highest radioactivity are the isotopes of U and Th and their daughter products. Another of common source of α-particles are the solder bumps that contain 210^{Po}. Solder bumps can cause higher SER if located in the close vicinity of the radiation-sensitive areas of the chip such as SRAM arrays. Beta radiation is usually not important from the SER standpoint because the ionization created by the β-particles is insufficient to cause a soft error in an SRAM cell.

Alpha particles induce the creation of large quantity of electron-hole pairs in the path of their strike. For an α-particle traveling in Si, an average of 3.6 eV of energy is lost for every electro-hole pair generated [118]. The rate of the energy loss increases when an α-particle is traveling in denser materials. As an α-particle is losing its velocity in a material, the ionization efficiency of an α-particle and hence the

Table 6.1 Alpha emissivity of some IC materials [118]

Material	Emissivity ($\alpha/\text{cm}^2-\text{h}$)
Eutectic Pb-based solders	$< 7.200 - < 0.002$
Mold compound	$< 0.024 - < 0.002$
30 μm of copper	< 0.002
Flip-chip underfill	$< 0.002 - < 0.001$
Fully processed wafers	< 0.001

electron-hole pair generation intensifies reaching the maximum near the end of the particle's trajectory. It is reported that over the travel range, an α-particle generates around 4-16 $fC/\mu m$ of charge per a micron of traveling through a material [118].

Alpha-radiating impurities can be found in some IC packaging materials, chemicals and materials used in the fabrication process of the chip. The emission can range widely depending of the quality and purification grade of the materials. A comparative table of α emissivity of some key production materials is shown in Table 6.1. As it is obvious from Table 6.1, α particles are primarily generated by the IC packaging materials and not by the materials used in the fabrication cycle.

6.3.2 Neutron-Induced ^{10}B Fission

Neutron-induced ^{10}B fission caused by low-energy ($\ll 1\,\text{MeV}$) is another major source of soft errors in deep sub-micron SRAMs [128]. Interaction of neutrons from ever-present cosmic rays and boron can generate enough charge to cause a soft error in areas of the IC containing boron. Boron is a common acceptor dopant that can be used in Si technologies for the formation of *p-type* regions by ion implantation and/or diffusion. Significant amounts of boron has also been used in boron-doped phosphosilicate glass (BPSG) blanket dielectric layers in technologies prior to 0.18 μm node [121]. While the BPSG dielectric helped lower reflow temperatures, better relief step coverage, improve alkali impurity gettering and reduce mechanical stress, it was also a significant source of soft errors. Boron is composed of two isotopes: ^{10}B ($\simeq 20\%$) and ^{11}B ($\simeq 80\%$) [118]. The isotope ^{10}B is unstable when bombarded with neutrons. Unlike most isotopes, the ^{10}B nucleus has 3 to 7 orders of magnitude larger neutron capture cross-section. When hit by a neutron, the ^{10}B nucleus breaks (fissions) into an α-particle, an excited ^{7}Li nucleus and a γ-photon. Both the α-particle and the ^{7}Li recoil are capable of creating soft errors in SRAMs, particularly in advanced technologies featuring low supply voltages.

Besides the α-particles, discussed in Section 6.3.1, ^{10}B fission produces a ^{7}Li recoil. Differential charge dQ/dx produced by ^{7}Li recoil with a kinetic energy of 0.84 MeV is estimated at 25 $fC/\mu m$ while the 1.47-MeV α-particle can generate only 16 $fC/\mu m$. While both the α-particle and the ^{7}Li recoil can create soft errors, the lithium recoil hitting a sensitive node in an SRAM cell has a higher probability of causing a soft error because it generates more charge per unit length of travel.

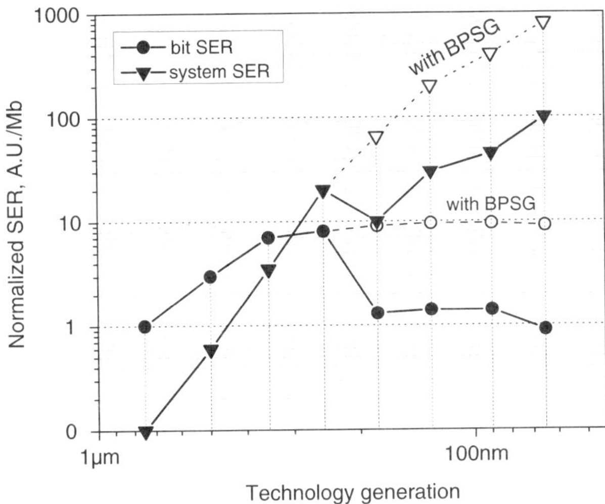

Fig. 6.3 SRAM single bit and SRAM system SER as a function of the technology generation. Note a drop in the SER levels beyond 250 nm technology generation due to the elimination of BPSG dielectric. Dotted lines show the simulated SER levels in case BPSG would have been used in the corresponding technology generations [121]

While neutrons with any energy level can cause ^{10}B fission, only low-energy neutrons need to be considered since the ^{10}B neutron capture cross-section rapidly decreases as the neutron energy increases [118]. Over 90% of ^{10}B fission reactions are caused by neutrons with energies below 15 eV. Only BPSG in close proximity to the Si substrate are reported to be a soft-error threat. ^{10}B fission is reported to be one of the dominant sources or even a primary cause of soft errors in BPSG-based processes [128]. When the BPSG was used as an inter-layer dielectric, the SRAM single bit SER was increasing with each technology generation [121]. When BPSG was discontinued at the technology nodes beyond 0.25 μm, the SER has saturated and even started to decrease.

SRAM devices with BPSG dielectric layers (0.25 μm-technology) were reported to exhibit nearly an order of magnitude greater SER than that of non-BPSG SRAMs (0.18 μm-technology) [128]. That clearly shows that the BPSG should not be used in applications requiring low SER. Figure 6.3 shows bit SER and system SER levels as a function of the technology generation. Note a significant drop in the SRAM bit SER beyond the 250 nm technology node due to the elimination of BPSG dielectric from the process flow. If BPSG usage was to be continued, the SER levels would have been as shown by the dotted lines in Figure 6.3.

6.3.3 High-Energy Cosmic Rays

Cosmic rays are another important source of radiation contributing to the SER. As cosmic rays enter Earth's atmosphere, they produce numerous other particles of the second and higher orders. Less than 1% of the primary flux reaches the sea level and the predominant particles include muons, protons, neutrons and pions [118]. Neutrons have relatively high flux and stability owing to their charge neutrality; hence are the main source of soft errors at the sea level. It is worth noting that the neutron flux is a function of the altitude and is increasing at higher altitudes. Flying at 10,000 ft increases the cosmic ray flux 10X.

Cosmic rays do not directly affect the SER but induce soft errors primarily by the neutron-induced silicon recoils. Neutrons collide with Si nuclei and transfer to them the energy that is sufficient to displace Si from the lattice. Typically, the Si nucleus breaks into fragments that generate charge that in turn causes soft errors. The detailed nature of this collisions is complex and not yet fully understood [129]. Si recoils are rapidly stopped by the lattice within a few microns from Si surface for most recoils.

Compared to the α-particles, the charge density per unit length generated by silicon recoils is significantly higher. Therefore, cosmic rays have higher potential for generating soft errors than α-particles. The charge density for neutron-generated Si recoils is 25–150fC/μm, while α-particles generate around 16 fC/μm [118].

6.4 Soft Errors and Defects in the Pull-Up Path of a Cell

Single Event Upsets (SEUs) are reported to induce voltage spikes on the storage nodes of SRAM cells that last about 100 ps [130]. The probability that an SRAM cell will recover from such a spike is a function of the current that drives the storage node. For instance, if a defect or a marginality exists in the pull-up current path of an SRAM cell, such a cell can be more susceptible to soft errors compared to a healthy cell with adequate pull-up current.

The term $I_{restore}\tau_{flip}$ in Equation 6.1 (Section 6.2) represents the restoring force or active feedback of an SRAM cell. Note that the $I_{restore}$ of the node storing a "1" is weaker than the $I_{restore}$ of the node storing a "0" due to the typical sizing ratio of the driver and the pull-up transistors $\frac{W_{driver}}{L_{driver}} / \frac{W_{driver}}{L_{pull-up}} \approx 1-2$ as well as the mobility ratio of electrons and holes $\mu_n/\mu_p \simeq 2-3$. Therefore, the Q_{crit} of the node storing a "1" is smaller than the Q_{crit} of the node storing a "0". That makes the node of an SRAM cell storing a "1" more susceptible to soft errors caused by the SEUs.

Now consider a cell with a resistive defect in the pull-up path. Resistive defects in the pull-up path similar to that presented in Figure 3.20 in Section 3.5.2 reduce $I_{restore}$ of the cell and thus, reduce Q_{crit}. Therefore, the presence of a non-catastrophic resistive defect in the pull-up path reduces the SNM and at the same time increases the probability of a soft error while remaining undetected by the regular memory test methods. Hence, the SRAM cell stability test techniques described in this book not only help to improve the test quality, but also sort out the parts that have potentially higher SER and are more susceptible to data loss due to ionizing radiation.

6.5 Soft Error Mitigation Techniques

Depending on the expected reliability of an electronic system, the same SER can be perceived by the customer either as being tolerable or unacceptable. For instance, CPUs with significant amounts of embedded SRAM can easily have the SER in excess of 50,000 FIT per chip, where $1\ FIT = 1\ error/10^9\ device\ hours$ [121]. Assuming that a CPU with the SER = 50,000 FIT runs around the clock, a soft error is likely to occur once every 2 years. If such a CPU is used in a cell phone and even if a soft error would corrupt a critical storage location, the customer would probably ignore the failure and his perception of quality of his cell phone would not be tarnished. However, when the same CPU is used in a mission-critical system or in a life-support system, one failure per chip every 2 years can become unacceptable. Moreover, if a hundred of such CPUs are used in a mainframe computer, the resulting SER of the system has to be multiplied by the number of CPUs. Simple calculations show that such a system will fail due to a soft error every week.

When a product's SER is deemed exceedingly high, SER mitigation techniques must be applied. To achieve acceptable radiation hardness of a device or a system, it may be necessary to employ one or more of the following SER mitigation techniques:

- Error Correction Codes (ECC)
- Elimination of radiation sources or reducing their intensity
- Radiation hardening by layout and circuit techniques
- Circuit and/or system redundancy

Error Correction Codes reduce the SER by correcting the corrupted data. Reducing or elimination of radiation sources will obviously also benefit the SER. Ultimately, any layout and/or circuit modification that increases Q_{crit} while maintaining or reducing Q_{coll} will help reduce the SER and make SRAM cells more radiation-tolerant. And finally, adding the redundancy on the circuit and/or system level is another powerful option that helps to reduce the sensitivity to ionizing radiation.

We will address these mitigation in more detail in the following sections.

6.5.1 *Error Detection and Correction*

A number of techniques were developed for detection and correction of errors in less-than-reliable storage elements such as the scaled-down SRAM cells. Errors generated in SRAM arrays by ionizing particles or stemming from the growing process variations and DPM levels of the scaled-down technologies can be detected and corrected with different degrees of success by a number of techniques. All of the detection and correction techniques used for SRAM arrays add a certain degree of redundancy into the system and therefore affect the system performance and occupy additional area. The choice of a detection/correction scheme is generally dictated by the required fault tolerance of the system.

Table 6.2 Error detection with a parity bit

Data word	With an even parity bit	With an odd parity bit
1111111	*1* 1111111	*0* 1111111
0000000	*0* 0000000	*1* 0000000
0001011	*1* 0001011	*0* 0001011
1011011	*0* 1011011	*1* 1011011

6.5.1.1 Parity Check

Parity check is the simplest error detection (but not an error correction) code. It works by adding an extra bit, the parity bit, to the data word so that the number of "1" data bits in the data word becomes even in case of even parity or odd in case of odd parity. Table 6.2 illustrates the concept of error detection with a parity bit.

The obvious advantage of the parity checking is its simplicity and minimal overhead. Parity check is effective in detecting an odd number of errors in a data word (including the parity bit). However, the generated error vector is not unique and does specify which bit/bits have been corrupted. When the number of corrupted bits in a data word is even, the parity bit will be still valid and the parity bit will not flag an error. Therefore, even number of corrupted bits in a word will not be detected by the parity check. Since the errors detected by the parity check are not unique and the location of the corrupted bits in the data word is unknown, the parity check is unable to correct the detected errors.

6.5.1.2 Error Correction Codes

In contrast to the parity error detection, the Error Correction Codes (ECC) add additional redundant bits that enable unique error vectors to be generated and the faulty bit location can be identified, provided that the redundant bits are arranged such that different corrupted bits in a data word produce different error results.

Error correction codes often use the concept of Hamming distance. Hamming distance between two binary data words w_1 and w_2 is defined as the number of the corresponding bit positions that are different and is equivalent to the number of "1"s in w_1 XOR w_2. For instance, the Hamming distance between $w_1 = 10010010$ and $w_2 = 11001010$ is 3. In the context of memory arrays, a non-zero Hamming distance signifies there is a difference (error) between the data written into the array and the data read back from the array. To be correctable, the Hamming distance between the sent (written) and the received (read back) data words should not be greater than one, i.e., the data word is allowed to contain only one error.

Hamming code general algorithm works by creating a code word that contains the initial data bits and the redundant parity check bits [131]. The code is putting the redundant parity bits into the bit positions that are the powers of two (bit positions #1,2,4,8, etc.). The other bit positions are reserved for the data that is to be encoded (bit positions #3,5,6,7,9,10, etc.). Each parity bit is used in calculating the parity of only certain bits in the coded data word. The position of each parity bit in a coded data word indicates the sequence of bits of the coded data word that this parity bit alternately checks or skips. For instance, a redundant parity check bit placed in a position n will skip $n - 1$ bits, check the next n bits, skip the next n bits, check the next n bits etc. The parity bit inserted into the code data word at the bit position # 2^x will be used to check the data bits that have the bit x set in their binary representation. Conversely, knowing the bit position, we can calculate the parity bits that will be used to check this bit. For instance, we want to check the bit #10 in a coded data word. Bit #10 in the binary form is expressed as 1010. Since the binary positions of the "1"s in 1010 correspond to $2^3 = 8$ and $2^1 = 2$, bit 10 will be checked by the parity bits in positions $8_{(10)} = 1000_{(2)}$ and $2_{(10)} = 0010_{(2)}$.

To further illustrate the principle of how Hamming codes are calculated and used to detect an error, let's consider an example when Hamming code is applied to a 16-bit data word "0011001100110011". Table 6.3 presents the principle of positioning the redundancy parity bits in a 16-bit data word. In this example **d1-d16** signify the original data bits and their positions in the resulting coded data word, while **p1-p5** signify the added parity bits and their positions in the resulting coded data word. Using the principle described in the above paragraph, we inserted the data bits into their appropriate positions signified by **bit#**. Using *even* parity, we calculated the

Table 6.3 Hamming(21,16) Code; correct 16-bit data word 0011001100110011 with 5-bit even parity information written into the memory array

bit#	d/p bits	*p1*	*p2*	*p3*	*p4*	*p5*	Without parity	With parity
1	***p1***	1						***1***
2	***p2***		***0***					***0***
3	d1	0	0				0	0
4	***p3***			***0***				***0***
5	d2	0		0			0	0
6	d3		1	1			1	1
7	d4	1	1	1			1	1
8	***p4***				***1***			***1***
9	d5	0			0		0	0
10	d6		0		0		0	0
11	d7	1	1		1		1	1
12	d8			1	1		1	1
13	d9	0		0	0		0	0
14	d10		0	0	0		0	0
15	d11	1	1	1	1		1	1
16	***p5***					***1***		***1***
17	d12	1				1	1	1
18	d13		0			0	0	0
19	d14	0	0			0	0	0
20	d15			1		1	1	1
21	d16	1		1		1	1	1

parity bits and inserted them into their respective positions. The second column of Table 6.3 shows the data and the parity bit positions in the resulting coded data word. A total of five parity bits is required to enable the error detection capability. Thus, the resulting coded data word will have 21 bits and the respective Hamming code can be denoted as Hamming(21,16) Code.

After the addition of the five redundant parity bits, the original 16-bit data word "0011001100110011" will become a 21-bit coded data word "10000111001100111-0011" (see the rightmost column of Table 6.3). Suppose now that after reading the data word back from a memory array, the final data bit **d16** in bit position #21 gets corrupted and flips from a "1" to a "0" (see Table 6.4). The new corrupted data word becomes "100001110011001110010". The corrupted bit **d16** will cause the failure of the parity check for the groups of the data bits which included bit **d16** in bit position #21. The "parity check" and the "parity bit" rows of Table 6.4 show the results of even parity check of when bit **d16** gets corrupted and the assigned parity bit values, respectively. We can see that the parity check will fail three times and each time the parity bit will be set to a "1".

The final step evaluates the value of the parity bits shown in Table 6.4. The bit with the lowest index (**p1**) is the least significant bit, and the bit with the highest index (**p5**) is the most significant bit, respectively. Table 6.5 shows the binary and the

Table 6.4 Hamming(21,16) Code; example of the corrupted data with bit in position #21 flipped from "1" to "0" that was read back from the memory array; even parity check failures are indicated by setting the parity bit to "1"

bit #	**d/p bits**	***p1***	***p2***	***p3***	***p4***	***p5***	**Without parity**	**With parity**
1	***p1***	***1***						***1***
2	***p2***		***0***					***0***
3	d1	0	0				0	0
4	***p3***			***0***				***0***
5	d2	0		0			0	0
6	d3		1	1			1	1
7	d4	1	1	1			1	1
8	***p4***				***1***			***1***
9	d5	0			0		0	0
10	d6		0		0		0	0
11	d7	1	1		1		1	1
12	d8			1	1		1	1
13	d9	0		0	0		0	0
14	d10		0	0	0		0	0
15	d11	1	1	1	1		1	1
16	***p5***					***1***		***1***
17	d12	1				1	1	1
18	d13		0			0	0	0
19	d14	0	0			0	0	0
20	d15			1		1	1	1
21	**d16**	**0**		**0**		**0**	**0**	**0**
Parity check:		*FAIL*	*PASS*	*FAIL*	*PASS*	*FAIL*		
Parity bit:		1	0	1	0	1		

Table 6.5 Determining the position of the faulty bit

	p5	*p4*	*p3*	*p2*	*p1*	Σ
Binary	1	0	1	0	1	
Decimal	16		4		1	*21*

decimal representation of the parity check bits resulted from checking the parity of the corrupted coded data word. The summing of the decimal values of the parity bits results in 21. This is signifying that the bit #21 in the coded data word is corrupted and needs to be flipped.

Flipping bit #21 reverts the corrupted coded data word "100001110011001110010" back to the original coded data word "100001110011001110011". The subsequent removal of the the Hamming codes results in the original data word of "0011001100110011".

Note that the parity bits do not check each other for corruption. Therefore, if a single parity bit check fails and of of the other parity checks succeed, then it is the parity bit itself that is corrupted and not any of the data bits which it checks.

In general, the number of detectable and correctable errors can be derived from the minimum Hamming distance d_{min} of an error correction code as $\lfloor \frac{d_{min}}{2} \rfloor$ and $\lfloor \frac{d_{min}-1}{2} \rfloor$, respectively. Successful detection and correction of a single error in a data word can be enabled by using Hamming codes with a minimum distance of three [132, 133], while a parity check has a minimum distance of two. However, with Hamming distance of three, a double-bit error will be indistinguishable from a different code with a single-bit error. Thus, a code with $d = 3$ can detect double-bit errors, but only if correction is not attempted. A Hamming code with the Hamming distance of four enables detection and correction of a single error and at the same time detection (but not correction) of a double error. This code is commonly called the Single Error Correction and Double Error Detection (SEC-DED) scheme.

Suppose two bits in positions a and b have the same bit at the 2^x position in their binary representations. Then they both will be checked by the same parity bit corresponding to position 2^x and so will remain the same. However, since $a \neq b$, some of the other parity checks will fail because they will differ in a and b and their parity bits will be altered. Thus, the Hamming code will be able to detect double-bit errors with a non-unique error signature and cannot distinguish them from single-bit errors.

Implementing the SEC-DED scheme on a 64-bit wide memory word requires the addition of eight extra parity bits to the code word. SEC-DED ECC scheme is effective to correct single-bit errors and can reduce of the perceived single-error SER that can be observed externally. A properly implemented SEC-DED scheme can virtually eliminate memory SER by reducing it by at least four orders of magnitude [134].

While double-bit and multiple-bit errors are not directly correctable by SEC-DED, the probability of a double error in a properly laid out SRAM arrays is rather small. Handling of multiple-bit errors can be achieved either by more powerful ECC schemes such as Double Error Correction and Triple Error Detection (DEC-TED) scheme. DEC-TED implementation requires more parity bits for achieving a larger Hamming distance and additional encoding/decoding circuitry. Double-bit errors can be caused by the ionizing particle strikes with an impact area spanning more than a single SRAM cell. Single-event upsets causing multiple single-bit errors are found to be rare. Maiz et al. [135] estimated the probability of two adjacent bits to fail within a single readout period as a result of two separate neutron events for an SRAM cell with eight adjacent neighboring cells as:

$$P = T_r \times F \times \frac{8}{bit\ count},$$

where T_r is the readout period and F is the average failure rate. For SRAM cells in 130 nm technology this probability was estimated to be $\sim 1.0 \times 10^{-7}$, which is negligibly small. Multiple bit failures are likely to be caused by neutrons. It was shown that α-particles did not cause multi-bit failures in 0.25 µm SRAM cells.

Another effective way to tackle multiple-bit errors is spacial separation of bits in the same correction word. This separation is commonly achieved by introduction of column (bit) interleaving. Column interleaving interleaves bits from different logic correction words in a single row. Since the bits from the same logic correction word in a column-interleaved design are physically further apart, a single strike of an ionizing particle cannot cause multiple-bit failures in the same logic correction word. Effectively, bit interleaving distributes the impact of a multi-bit soft error in a single logic correction word into multiple single-bit soft errors on multiple logic correction words. Degalahal et al. [136] showed that the number of double-bit errors in a drowsy cache without column interleaving can contribute up to 1.6% of the number of single-bit errors. Whereas, when column interleaving is implemented, the number of double-bit and multi-bit errors reported to become zero. Column interleaving rules specify the minimum physical distance between the bits that belong to the same logic correction word. The minimum acceptable bit separation for a given technology and cell layout is defined based on the multiple bit failure probabilities [135], the sensitivity of the cells and the actual cell size. In error-corrected SRAMs, the typical physical spacing between bits from the same logic correction word is at least four to eight bits [121]. Column interleaving therefore alleviates the area, power and performance burden of multi-bit ECC schemes and makes the single error correction an adequate means against soft errors. An additional benefit of column interleaving is the area and power savings derived from sharing of the same read and write circuitry among a number of interleaved columns.

While applying ECC to the data stored in SRAM arrays and column interleaving results in significantly lower SER, the SRAM cells may not be the only contributors to the system's SER. The scaling trends shown in Figure 6.4 indicate that the SER in a system that contains SRAM arrays with ECC and the periphery sequential gates can become limited by the SER contributed by the sequential gates that latch in the soft errors. The overall system-level SER to the first order can be modeled as the sum of the contributions from both the memory and the logic components [137]:

$$SER_{sys} = \delta_{mem} SER_{mem} + \delta_{logic} SER_{logic},$$

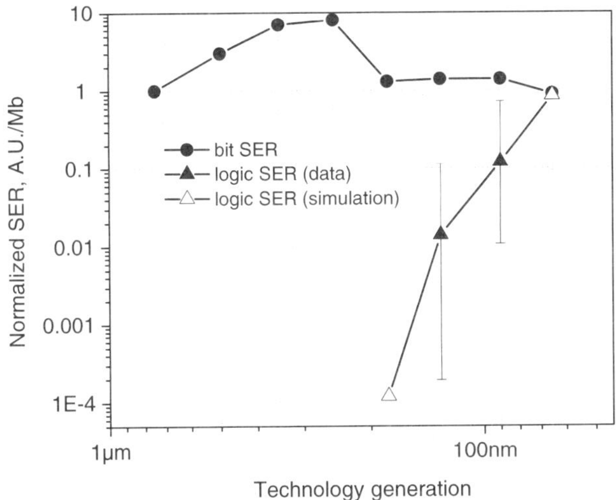

Fig. 6.4 The scaling trends of the SRAM bit SER and the flip-flop/latch SER. Data obtained from multiple sources including test structures, product characterizations and simulations. SEC-DEC error correction reduces the SRAM bit SER by approximately four orders of magnitude. Large error bars shown in the logic SER trend are due to multiple types of sequential circuits tested [123]

where δ_{mem} and δ_{logic} are the effective derating factors of the system memory and logic SER. They represent the fraction of soft errors that are observed on the system level. The derating factors depend on the system architecture, chip layout susceptibility to soft errors and the application being run on the system [138]. Zhu et al. [139] confirm the validity of using the sum of components SER to compute the system SER with an appropriate memory usage derating factor.

Growing SER of the periphery circuits can effectively limit the efficacy of the ECC schemes to keep the system SER at an acceptable level. This trend is disturbing since even at the 0.13 μm node SER in the sequential logic are high enough to limit the efficacy of ECC since the SER of the logic is only 100–1000X lower than SRAM bit SER, while the failure rate of SRAM arrays with ECC is at least ten times lower [134]. Further experiments are required to determine whether the logic SER will continue to grow exponentially [140] or will saturate similar to the SRAM SER. If the failure rate of the sequential logic in a certain product is too high, measures should be taken to reduce the logic SER to a level similar to the SER of ECC-protected SRAM. Ensuring low SER in critical systems may require finding the sensitive flip-flop locations in critical data paths and applying fault-tolerant techniques such as triple node redundancy with a majority voter that we will discuss in more detail in Section 6.5.1.3.

Another possible use of the ECC scheme in SRAM mitigates the effect of cells with marginal stability on the minimal operating voltage of the array [141]. Suppose an SRAM is provided with a SEC-DED ECC scheme which can correct one error in a cache line and detect two errors without correction. Then, if for any reason, be it a single-event upset or a low-voltage operation, one of the cells in the cache line flips its data value, it can be corrected by the error correction circuit.

Statistically, large SRAM arrays are more likely to contain cells with marginal stability that may start failing at a slightly higher supply voltage than the rest of the cell population in the array. These unstable cells may be randomly distributed throughout the array. By applying an error correction scheme it is possible to reduce the array voltage to a lower value which is more characteristic of the rest of the cell in the array and rely on the ECC circuitry to correct the small number of cells that would fail at a reduced voltage. Because only one row of the SRAM array is accessed at a time, ECC can effectively correct single failing cells belonging to the same column. Indeed, in this case only one error will be present in a cache line and the errors in other rows will not be interfering with the error correction. The effectiveness of the SEC-DEC scheme is, however, limited to only single failing bit in a cache line and ECC schemes can be ineffective for correcting multiple-bit failures. Even applying the more elaborate DEC-TED ECC scheme one can correct only two errors in a cache line at the expense of the additional circuitry and the associated area overhead. However, if more than two bits with marginal stability are detected to be failing in the same cache line, even the DEC-TED scheme is not sufficient to correct the errors. Repairing the faulty row by replacing it with a redundant row can be a more efficient means for tackling the multiple-bit failures in the same cache word.

6.5.1.3 Circuit and/or System Redundancy

Circuit and/or system redundancy are the most reliable methods to bring the system SER to near-zero levels. Although the redundancy methods are the most effective, they are also the most expensive in terms of area and performance. Redundant systems are often employed in highly-reliable real-time applications such as life-critical missions, aircraft and space apparatus control and transactional processing [142].

In system and/or circuit redundancy two or multiple identical system or circuit components run in lock-step executing the same code in the same time. A mismatch in the data detected in a dual-component system will result in a restart of the system. When two or more redundant components are used, a majority voter is used to identify which of the data outputs supply the correct data. This architecture allows a soft error in a single signal path to be ignored in favor of the others, or the majority, data that is assumed to be correct. Thus, the correct data "wins" the vote and appears at the output which is passed to the next stage.

The downside of the circuit/system redundancy is the extra area, power, latency and delay which is inherent to the redundant schemes. For instance, a triple redundancy scheme conceptually shown in Figure 6.5 would consume approximately triple the chip area in addition to the majority voter logic. The TMR circuit scheme is often called a spacial-multiplexed scheme. The layout of the redundant circuits typically minimizes the probability of an SEU that induced enough charge to disrupt two or more of the redundant circuits simultaneously [143].

Time-multiplexed designs sample input at several different times and a voter circuit sets the output on the basis of matching inputs. The effectiveness of the

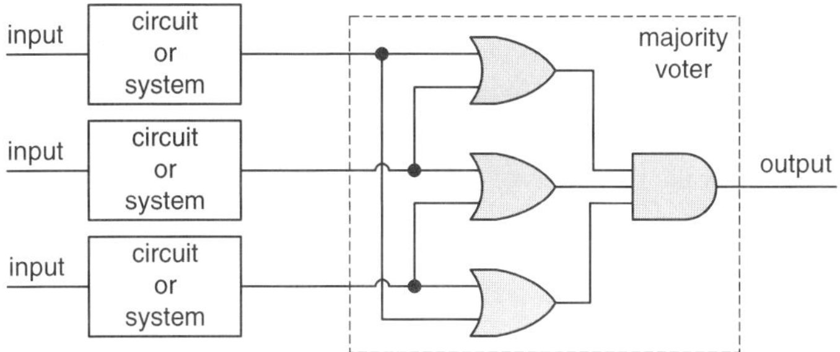

Fig. 6.5 Conceptual diagram of the Triple Modular Redundancy (TMR) fault tolerant scheme

time-multiplexed scheme is based on the fact that the probability of two independent errors occurring in the same circuit path within a small time period is exceedingly low.

A combination of time- and spatial-multiplexed SER mitigation techniques can improve the soft error tolerance even further at the expense of larger area and complexity.

6.5.2 Removing Radiation Sources or Reducing Their Intensity

Soft errors caused by the α particles can be mitigated in several ways. One option pertains to the purity of the materials used in the manufacturing process and packaging of the chip. Careful choice of solders and mold and underfill compounds with reduced emissivity rates is important to mitigate the SER created by α particles. The best materials as well as the detection sensitivity are limited today at $0.001\alpha/\text{cm}^2 - \text{h}$. Therefore, using highly purified materials with emissivity $0.001\alpha/\text{cm}^2 - \text{h}$ will keep the α-related SER below the measurable limit. Another important consideration in mitigating the α-related SER is to keep local sources of α-radiation such as solder bumps away from high-sensitivity areas of the chip (e.g., SRAM arrays). Keeping the packaging components with high emissions from the sensitive regions of the chip is accomplished by special design rules and is used in flip-chip packaging. However, if a chip has many small SRAM arrays that are sensitive to α radiation, then the effectiveness of solder bump allocation diminishes. And lastly, coating the chip with thick polyimide film also has been reported to reduce the α-related SER [118]. This mitigation technique, however, is not feasible for flip-chip packaging, but other packages can benefit from coating the chips prior to bonding and packaging.

The most effective method of mitigating the SER caused by the neutron-induced ^{10}B fission eliminates the BPSG from the semiconductor process flow. The neutron-induced α-particles and the emitted ^{7}Li recoils during the ^{10}B fission have limited range. Therefore, the lowest BPSG layers are the most critical for the SER mitigation caused by the neutron-induced ^{10}B fission. The SER can be mitigated by replacing the lowest level of BPSG with a ^{10}B-free inter-layer dielectric. Another option uses ^{11}B-enriched BPSG process without sacrificing the benefits of its physical and chemical properties or without the need to update the equipment and modify technology steps [118].

The measures for the reduction of SER that work for α-particles are not effective to improve the neutron-induced SER. While concrete attenuates the neutron flux, the concrete shielding is only feasible for mainframes and large computing systems. In terms of the shielding, not much can be done for other applications such as desktops and mobile devices. Therefore, the neutron-induced SER can only be improved by reducing the SRAM sensitivity to soft errors by implementing special design and process solutions, some of which will be described in the next sections.

Although large SER reductions are possible either by removing or shielding radiation source of α-particles, a large portion of the high-energy cosmic neutrons will always reach the sensitive SRAM arrays and cause soft errors. Therefore, while accurate estimation of the cumulative SER of a product requires to account for all three major sources of soft errors, the high-energy cosmic neutron radiation is what ultimately defines and limits the SER of a product.

6.5.3 Increasing the Capacitance of the Storage Nodes

An obvious way to improve the SER of an SRAM cell is to increase the capacitance of its storage nodes. Note that the SER can only be reduced when adding the gate capacitance and/or the interconnect capacitance. The same cannot be said about the junction diffusion capacitance. Adding more diffusion capacitance by increasing the node area can in fact increase the charge collection efficiency of the sensitive node during a strike by an ionizing particle. So, the effect of adding more diffusion capacitance can be opposite to the expected and can increase the SER.

As was shown above in Equation 6.1, the critical charge Q_{crit} is proportional to $C_{node} \times V_{node}$. Therefore, maximizing either the storage node capacitance or voltage will result in a larger Q_{crit}. Liaw et al. [144] proposed adding extra capacitance to the storage nodes of an SRAM cell. Figure 6.6 schematically illustrates this idea. Such capacitors can be formed by using the parasitic capacitance between the interconnect metal layers, e.g., the first and the second levels of metallization. Connecting extra capacitance to the storage nodes of an SRAM cell effectively increases the C_{node} capacitance. As a result, the Q_{crit} of a cell that is modified this way will be larger. Therefore, upsetting a storage node in such a cell might require particles with higher energy and/or more sustained charge introduced by an α-particles or neutrons. The longer discharge time of the more capacitive storage node will lead to a decrease in the SER.

ST Microelectronics reported using vertical Metal-Insulator-Metal capacitors in SRAM cells for SER reduction [145]. The capacitors, shaped like tall cylinders, are located in the intermediate layers of interconnect between M1 and the top Metal layer. Since there is no SRAM cell interconnect in those levels, the capacitors don't

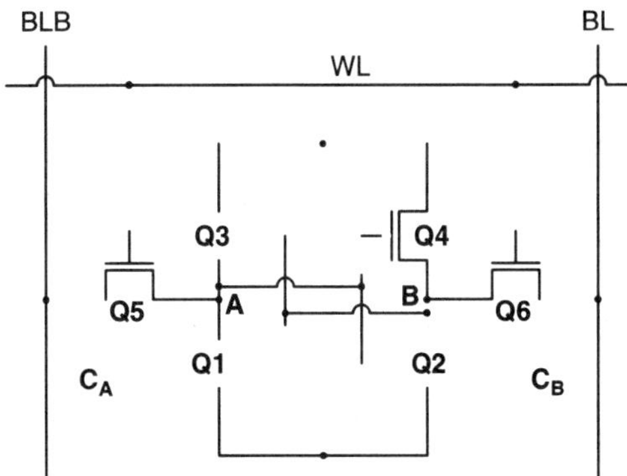

Fig. 6.6 Capacitors C_A and C_B connected to the storage nodes A and B help to improve the SER of an SRAM cell [144]

get in the way of the cell or increase its area, they just take up unused space above it. Of course, that limits the ability to route over SRAM resulting in about a 5% area penalty. The company fabricated a 120 nm test chip that included conventional SRAM arrays and arrays of the new radiation-hardened cells. The chips were bombarded with alpha particles and with neutrons. The hardened cells showed a 250x improvement in the SER over the standard ones. ST has reported $SER < 10$ FIT/Mb during 1.2 V operation and no failures at 1.32 V.

6.5.4 Inserting Resistors in the Feedback Loop of an SRAM Cell

Some applications such as Field Programmable Grid Arrays (FPGA) allow more flexibility in the cell-level radiation hardening. SRAM cells in FPGA applications are used mainly in the read mode and are usually written only once during the power-up sequence to define the FPGA configuration. In most FPGA applications the data stored in the configuration SRAM are never changed after the power-up is completed. Therefore, radiation-hardening cell-level techniques that affect SRAM cell access speed are more acceptable in FPGA SRAM arrays. High-energy neutrons, generated in the upper atmosphere, can strike a configuration cell of an SRAM FPGA. The energy of the collision can change the state of the configuration SRAM cell and thus change the logic, routing, or I/O behavior in an unpredictable way. The consequence of this type of error can be a complete system failure [146]. Therefore, radiation hardening of SRAM configuration arrays in FPGAs is extremely important. For instance, SRAM arrays that are used in FPGA for configuration data storage can be hardened against radiation using large-value resistors [147]. Figure 6.7 illustrates this concept.

Compared to a regular 6T SRAM cell, the SRAM cell shown in Figure 6.7 contains two extra resistors: $R1$ and $R2$. Another difference is the input capacitances $C1$ and $C2$ of inverters $Q1-Q3$ and $Q2-Q4$ respectively that is shown in Figure 6.7.

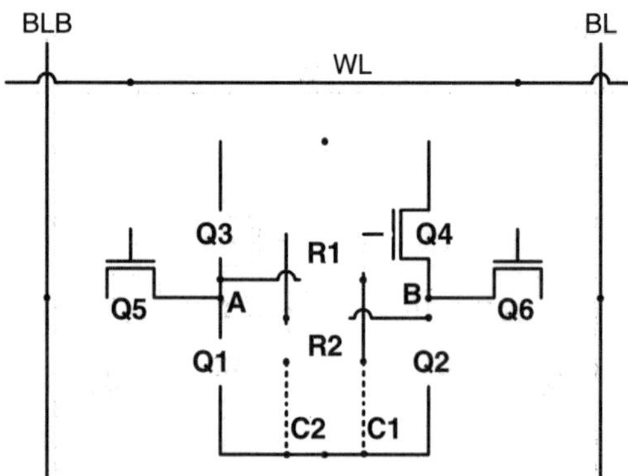

Fig. 6.7 Resistors $R1$ and $R2$ are inserted in the feedback loop of an SRAM cell to increase the RC time constant τ of the cell [147]

Together with resistors $R1$ and $R2$, capacitors $C1$ and $C2$ form RC chains $R1C1$ and $R2C2$ with RC time constants τ_1 and τ_2. With sufficiently high resistance of $R1$ and $R2$, the time constants τ_1 and τ_2 will be significantly larger than those of a regular SRAM cell. McCollum [147] suggests using $R1$ and $R2$ with resistance of the order of several MΩ. Note that when resistors $R1$ and $R2$ are introduced in the SRAM cell (Figure 6.7), the input of the inverter $Q1-Q3$ is no longer directly connected to the output of the inverter $Q2-Q4$ and vice versa. Therefore, the diffusion region of node B is not directly connected to the input of the inverter $Q1-Q3$ as well as the diffusion region of node A is not directly connected to the input of the inverter $Q2-Q4$. Suppose a high-energy particle hits node A that is storing a "1". Then the charge carriers generated by the particle and collected by node A will create a current spike on node A. However, thanks to a high-ohmic resistor $R1$ and capacitor $C1$, the RC constant τ_1 will slow down the voltage transition on the input of the inverter $Q2-Q4$. In order to upset the state stored in such an SRAM cell, the charge generated by the radiation hit must create a disturbance exceeding the SNM of the SRAM cell. This slowly growing disturbance will be applied to the capacitor $C1$ through the resistor $R1$. The disturbance caused by a high-energy particle hit on node A will reach the input of the inverter $Q1-Q3$ after a certain *recovery time* elapses. The SRAM cell shown in Figure 6.7 will not be flipped by an ionizing particle as long as the RC time constant of the cell exceeds the recovery time of the output of the inverter. The amount of the recovery time of such a radiation-hardened SRAM cell can be adjusted by varying the resistance of resistors $R1$ and $R2$.

Introducing resistors $R1$ and $R2$ inevitably increases the write time of an SRAM cell. However, since the SRAM configuration arrays in FPGAs are written only once during the power-up, the increase in the write time can be tolerated. To ensure reliable write operation, the write time for such a radiation-hardened SRAM cell should exceed the RC time constant. For instance, if the recovery time is 10 ns and the inverter input capacitance is 5 fF, then the resistor should be about 2 MΩ. Resistors $R1$ and $R2$ can be implemented in poly-Si with no diffusion region that can collect the charge caused by the particle hit [147].

6.6 Leakage-Reduction Techniques and the SER

Constant electric field scaling adopted for the recent semiconductor technologies as well as the drive to reduce the dynamic power of the chips necessitate scaling down the supply voltage. The threshold voltages are following the same trend to maintain the performance requirement with every new technology generation. While smaller threshold voltages result in exponentially higher sub-threshold leakage, the lower power supply voltages and the shrinking node capacitances increase the SER.

In the SER estimation method developed by Hazucha et al. [148], the SER exhibits an *exponential* dependence on the Q_{crit} [149]:

$$SER \propto N_{flux} \times CS \times exp(\frac{-Q_{crit}}{Q_S}),$$

where N_{flux} is the intensity of the neutron flux, CS is the area of the cross-section of the node and Q_S is the charge collection efficiency which depends on doping concentration and profile. The critical charge is reported to be in almost linear relation to the supply voltage. Figure 6.8 shows the Q_{crit} and the leakage energy savings per cycle as a function of the supply voltage [136]. Figure 6.8 shows that both the leakage energy and Q_{crit} are proportional to the power supply voltage V_{DD}. All other factor being equal, the reduction of the Q_{crit} at the lower supply voltages leads to an exponential increase in the SER. We demonstrated that lowering the supply voltage also reduces the SNM of an SRAM cell (Section 3.5.3). Thus, it is important to balance the leakage saving benefits with reliability concerns when choosing the minimal supply voltage of an SRAM array V_{DD_min}.

Since the area of SRAM arrays can dominate the total chip area, the SRAM leakage power can be a significant part of the total chip leakage. Therefore, leakage control techniques are common in large SRAM arrays. A number of leakage reduction techniques works by reducing the power supply voltage of the array when it is not accessed [113, 149, 150]. The reduced power supply voltage in sleep mode has to be carefully chosen to be high enough for reliable data retention. Such techniques

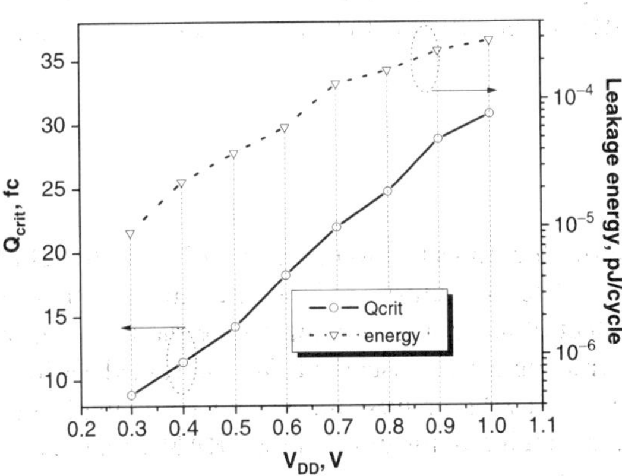

Fig. 6.8 Q_{crit} and the leakage energy savings per cycle as a function of the supply voltage [136]

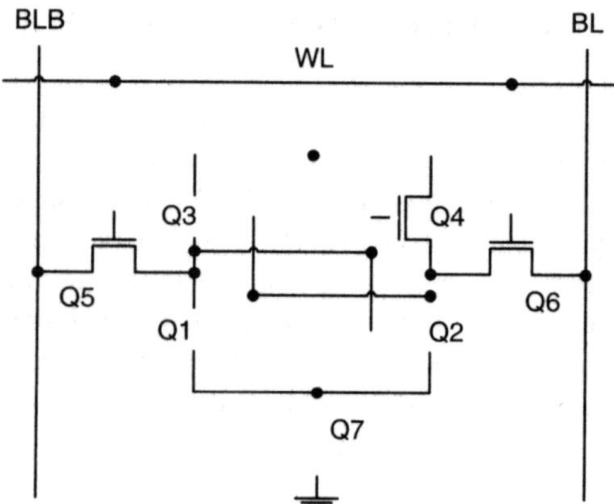

Fig. 6.9 Data-retention gated-ground SRAM cell [149]

can significantly cut down on the leakage power consumption. Up to 5X leakage reduction is reported [113]. Large cache arrays typically feature ECC circuitry. Error correction is necessary to suppress the negative impact of the leakage power reduction techniques on the SER to acceptable levels.

The schematic of one of the SRAM cell leakage saving techniques, the Data-Retention Gated-ground (DRG) [149], is shown in Figure 6.9. When a cache with the DRG leakage saving technique is not accessed, i.e., when its $WL = 0$, the footer transistor $Q7$ is also off. Effectively, the conducting path from the node storing a "0" to the ground in a non-accessed cells is cut off by transistor $Q7$. Proper sizing and an optimum value of V_{TH} may be required in order to control the virtual ground (the V_{DS} of $Q7$) and to ensure that the DRG cache will reliably store its data in the leakage-reduction mode.

Degalahal et al. [126] conducted a comparative analysis quantifying the critical charge of the DRG vs. a regular 6T SRAM cell. For a $1 \rightarrow 0$ transition, the critical charge Q_{crit} of a regular 6T SRAM cell is reported to be around 25% larger than that of the same SRAM cell utilizing the gated-ground leakage control technique.

6.7 Summary

This chapter reviewed the impact of ionizing particles on the reliability of SRAM circuits and systems. The mechanism and the three major sources of soft errors in SRAMs and the effect of technology scaling on the radiation sensitivity of SRAM arrays and peripheral logic was discussed. We presented various approaches to radiation hardening of SRAM circuits and systems including limiting the radiation exposure, layout and circuit modifications of the SRAM cell, radiation tolerant array layout techniques, error detection and correction and finally, the circuit/system redundancy. We showed that poor SNM due to defects in the pull-up path of an SRAM cell can compromise the cell soft error rate. Leakage control techniques exploiting the exponential leakage reduction at reduced supply voltages are shown to be another factor that can impact the data integrity of SRAM cells exposed to radiation sources.

References

1. A. Meixner and J. Banik. Weak write test mode: An SRAM cell stability design for test technique. In *Proc. IEEE International Test Conference (ITC)*, pages 1043–1052, November 1997.
2. D.-M. Kwai, H.-W. Chang, H.-J. Liao, C.-H. Chiao, and Y.-F. Chou. Detection of SRAM cell stability by lowering array supply voltage. In *Proc. of the Ninth Asian Test Symposium (ATS 2000)*, pages 268–273, December 2000.
3. C. Kuo, T. Toms, B.T. Neel, J. Jelemensky, E.A. Carter, and P. Smith. Soft-defect detection (SDD) technique for a high-reliability CMOS SRAM. *IEEE Journal of Solid-State Circuits (JSSC)*, 25:61–67, February 1990.
4. ITRS. International Technology Roadmap for Semiconductors - 2003 (ITRS-2003); *http : //public.itrs.net/*.
5. ITRS. International Technology Roadmap for Semiconductors (ITRS-2004) update; *http : //www.itrs.net/common/2004update/2004update.htm*.
6. A. Pavlov, M. Sachdev, and J. Pineda de Gyvez. An SRAM weak cell fault model and a DFT technique with a programmable detection threshold. In *Proc. IEEE International Test Conference (ITC)*, pages 1106–1115, November 2004.
7. J. Wuu, D. Weiss, C. Morganti, and M. Dreesen. The asynchrounous 24MB on-chip level-3 cache for a dual-core Itanium-family processor. In *IEEE International Solid-State Circuits Conference*, pages 488–489, April 2005.
8. B.F. Cockburn, F. Lombardi, and F.J. Meyer. DRAM architectire and testing. *IEEE Design & Test of Computers*, pages 19–21, January 1999.
9. A. Sharma. *Advanced Semiconductor Memories: Architectures, Designs and Applications*. Wiley Inter-Science, 2003.
10. G. Moore. Cramming more components onto integrated circuits. *Electronics*, 38:114–117, October 1965.
11. G.E. Moore. Progress in digital integrated electronics. In *IEEE Electron Device Meeting*, pages 11–13, October 1975.
12. D. Grier. The innovation curve. *Computer*, 39:8–10, February 2006.
13. B. Chappel and I. Young. V_{DD} modulated SRAM for highly scaled, high performance cache. *US Patent No. 6,556,471*, January 2003.
14. M. Bushnell and V. Agrawal. *Essentials of electronic Tesing for Digital, Memory and Mixed-Signal VLSI Circuits*. Kluwer, 2000.
15. C. Stroud. *A Designer's Guide to Built-In Self-Test*. Kluwer Academic Publishers, 2002.
16. J. Pineda de Gyvez and D. Pradhan. *Integrated Circuit Manufacturability*. Institute of Electrical and Computer Engineers Inc., 1998.
17. T. Harazsti. *CMOS Memory Circuits*. Kluwer, 2000.
18. Sizing Cu-11 dense SRAM and fuse redundancy options. IBM application note. *http : //www – 306.ibm.com/chips/techlib/techlib.nsf/techdocs/*, 2005.

19. S. Naffziger, T. Grutkowski, and B. Stackhouse. The Implementation of a 2-core, Multi-Threaded Itanium-Family Processor. In *IEEE International Solid-State Circuits Conference (ISSCC)*, pages 182–184, February 2005.
20. L. Kim and R. Dutton. Metastability of CMOS latch/flip-flop. *IEEE Journal of Solid-State Circuits (JSSC)*, 25:942–951, August 1990.
21. J. Rabaey, A. Chandrakasan, and B. Nicoloc. *Digital Integrated Circuits: A Design Prospective. Second Edition*. Prentice Hall, 2003.
22. N. Lu, L. Gerzberg, and J. Meindl. Scaling limitations of monolithic polycrystalline-silicon resistors in VLSI Static RAM's and logic. *IEEE Journal of Solid State Circuits (JSSC)*, SC-17:312–320, April 1982.
23. K. Zhang, U. Bhattacharya, Z. Chen, F. Hamzaoglu, D. Murray, N. Vallepali, Y. Wang, B. Zheng, and M. Bohr. A 3-GHz 70-Mb SRAM in 65-nm CMOS technology with integrated column-based dynamic power supply. *IEEE Journal of Solid State Circuits (JSSC)*, 41:146–151, January 2006.
24. K. Imai et al. A 0.13μm CMOS technology integrating high-speed and low power/high density devices with two different well/channel structures. In *IEEE International Electron Devices Meeting IEDM Technical Digest*, pages 667–690, October 1999.
25. NEC. In *ISSCC Digest of Technical Papers*, February 2001.
26. S. Masuoka et al. A 0.99-m^2 loadless four-transistor SRAM cell in 0.13μm generation CMOS technology. In *Digest of Technical Papers of Symposium on VLSI Technology*, pages 164–165, June 2000.
27. K. Takeda et al. A 16-Mb 400-MHz loadless CMOS four-transistor SRAM macro. *IEEE Journal of Solid-State Circuits (JSSC)*, 35:1631–1640, November 2000.
28. C. Lage, J. Hayden, and C. Subramanian. Advanced SRAM technology – the race between 4T and 6T cells. In *IEEE International Electron Devices Meeting IEDM Technical Digest*, pages 271–274, December 1996.
29. K. Noda, K. Matsui, K. Takeda, and N. Nakamura. A loadless CMOS four-transist SRAM cell in a 0.18-μm logic technology. In *IEEE Transactions on Electron Devices*, pages 2851–2855, December 2001.
30. P. Bai et al. A 65nm logic technology featuring 35nm gate lengths, enchanced channel strain, 8 Cu interconnect layers, low-k ILD and 0.57μm^2 SRAM cell. In *IEEE International Electron Devices Meeting IEDM Technical Digest*, pages 657–660, December 2004.
31. F. Arnaud et al. Low cost 65nm CMOS platform for low power and general purpose applications. In *Symposium on VLSI Technology Digest of Technical Papers*, pages 10–11, June 2004.
32. F. Boeuf et al. 0.248μm^2 and 0.334μm^2 conventional bulk 6T-SRAM bit cells for 45nm node low cost – general purpose applications. In *Symposium on VLSI Technology Digest of Technical Papers*, pages 130–131, June 2005.
33. M. Yamaoka, K. Osada, and K. Ishibashi. 0.4V Logic-library-friendly SRAM array using rectangular-diffusion cell and delta-boosted-array voltage scheme. Pages 934–940, June 2004.
34. R. Venkatraman, R. Castagnetti, O. Kobozeva, F. Duan, A. Kamath, S. Sabbagh, M. Vilchis-Cruz, J. Jhu Liaw, Jyh-Cheng You, and S. Ramesh. The design, analysis, and development of highly manufacturable 6T SRAM bitscells for SoC applications. *IEEE Transactions of Electron Devices*, 52:218–226, February 2005.
35. B. Cheng, S. Roy, and A. Asenov. The impact of random doping effects on CMOS SRAM cell. In *Proc. IEEE Solid-State Circuits Conference ESSCIRC*, pages 219–222, Leuven, Belgium, 2004.
36. T. Hirose et al. A 20-ns 4-Mb CMOS SRAM with hierarchical word decoding architecture. *IEEE Journal of Solid-State Circuits (JSSC)*, 25:1068–1073, October 1990.
37. B. Amurtur and M. Horowitz. A replica technique for wordline and sense control in low-power SRAMs. *IEEE Journal of Solid-State Circuits (JSSC)*, 33:1208–1219, August 1998.
38. M. Eisele et al. The impact of intra-die device parameter variations on path delays on the design for yield of low voltage digital circuits. In *IEEE International Symposium Low Power Electronic Design*, pages 237–242, October 1996.

39. S. Tachibana et al. A 2.6-ns wave-pipelined CMOS SRAM with dual-sensing-latch circuits. *IEEE Journal of Solid-State Circuits (JSSC)*, 30:487–490, April 1995.
40. S. Schuster et al. A 15-ns CMOS 64K RAM. *IEEE Journal of Solid-State Circuits (JSSC)*, 21:704–711, October 1986.
41. A. Bhavnagarwala, X. Tang, and J. Meindl. The impact of intrinsic device fluctuations on CMOS SRAM cell stability. *IEEE Journal of Solid-State Circuits (JSSC)*, 36:658–665, April 2001.
42. F. Yang, J. Hwang, and Y. Li. Electrical characteristic fluctuations in sub-45 nm CMOS devices. In *Proc. of IEEE Custom Integrated Circuits Conference CICC*, pages 691–694, September 2006.
43. F. Lai and C. Lee. On-chip voltage down converter to improve SRAM read-write margin and static power for sub-nano CMOS technology. *IEEE Journal of Solid-State Circuits (JSSC)*, 42:2061–2070, September 2007.
44. C. Hill. Definitions of noise margin in logic systems. *Mullard Tech. Commun.*, 89:239–245, September 1967.
45. J. Lohstroh. Static and dynamic noise margins of logic circuits. *IEEE Journal of Solid-State Circuits (JSSC)*, SC-14:591–598, June 1979.
46. J. Lohstroh, E. Seevinck, and J. de Groot. Worst-case static noise margin criteria for logic circuits and their mathematical equivalence. *IEEE Journal of Solid-State Circuits (JSSC)*, SC-18:803–807, December 1983.
47. T. DeMassa and Z. Ciccone. *Digital Integrated Circuits*. Wiley, 1996.
48. S. Mitra. *An Introduction to Digital and Analog Integrated Circuits and Applications*. Harper & Row, 1980.
49. L. Glasser and D. Dobberpuhl. *The Design and Analysis of VLSI Circuits*. Addison-Welsley, 1985.
50. J. Hauser. Noise margin criteria for digital logic circuits. *IEEE Transactions on Education*, 36:363–368, November 1993.
51. E. Seevinck, F. List, and J. Lohstroh. Static-noise margin analysis of MOS SRAM cells. *IEEE Journal of Solid-State Circuits (JSSC)*, SC-22:748–754, October 1987.
52. O. Semenov, A. Vassighi, and M. Sachdev. Impact of technology scaling on thermal behavior of leakage current in sub-quarter micron MOSFETs: perspective of low temperature current testing. *Microelectronics Journal*, pages 985–994, October 2002.
53. T. Sakurai and A.R. Newton. Alpha-power law MOSFET model and its applications to CMOS inverter delay and other formulas. *IEEE Journal of Solid-State Circuits (JSSC)*, 25:584–594, April 1990.
54. S. Vemuru, N. Scheinberg, and E. Smith. Short-circuit power dissipation formulae for CMOS gates. In *IEEE International Symposium on Circuits and Systems (ISCAS)*, pages 1333–1336, May 1993.
55. TSMC. Taiwan Semiconductor Manufacturing Company literature; *http://www.tsmc.com/english/function/f07.htm.*
56. P.A. Stolk, H.P. Tuinhout, et al. CMOS device optimization for mixed-signal technologies. In *IEEE International Electron Devices Meeting IEDM Technical Digest*, pages 10.2.1–10.2.4, October 2001.
57. A. Asenov, A.R. Brown, J.H. Davies, S. Kaya, and G. Slavcheva. Simulation of intrinsic parameter fluctuations in decananometer and nanometer-scale MOSFETs. *IEEE Transactions on Electron Devices*, 50:1837–1852, 2003.
58. F.J. Ferguson and J. Shen. Extraction and simulation of realistic CMOS faults using inductive fault analysis. In *IEEE International Test Conference*, pages 475–484, November 1988.
59. Carafe inductive fault analysis IFA tool; *http://sctest.cse.ucsc.edu/carafe/.*
60. A. van de Goor. *Testing Semiconductor Memories: Theory and Practice*. A. van de Goor, 2001.
61. A. van de Goor. An industrial evaluation of DRAM tests. *IEEE Design & Test of Computers*, pages 430–440, September 2004.
62. S. Hamdioui, R. Wadsworth, J. Reyes, and A. van de Goor. Importance of dynamic faults for new SRAM technologies. In *IEEE European Test Workshop*, May 2003.

63. S. Hamdioui, Z. Al-ars, and A. van de Goor. Testing static and dynamic faults in random access memories. In *IEEE VLSI Test Symposium*, pages 395–400, October 2002.
64. A. Pavlov. *Design and Test of Embedded SRAMs*. PhD thesis, University of Waterloo, Canada, 2005.
65. M. Sachdev, V. Zieren, and P. Janssen. Defect detection with transient current testing and its potential for deep submicron CMOS ICs. In *IEEE International Test Conference*, pages 204–213, October 1998.
66. R. Dekker, F. Beenker, and L. Thijssen. A realistic fault model and test algorithms for static random access memories. *IEEE Transactions of Computer-Aided Design*, 9:567–572, June 1990.
67. W. Maly. Modeling of lithography related yield losses for CAD of VLSI circuits. *IEEE Transactions of Computer-Aided Design*, 4:166–177, July 1985.
68. T. Saito, H. Ashihara, K. Ishikawa, M. Miyauchi, Y. Yamada, and H. Nakano. A reliability study of barrier-metal-clad copper interconnects with self-aligned metallic caps. *IEEE Transactions on Electron Devices*, 51:2129–2135, December 2004.
69. R. Montanés, J. Pineda de Gyvez, and P. Volf. Resistance characterisation of open defects. *IEEE Design and Test of Computers*, 19:18–26, September 2002.
70. E. Ogawa et al. Stress-induced voiding under vias connected to wide Cu metal leads. In *Proc. of IRPS*, page 312–321, October 2002.
71. D. Adams, R. Abbott, X. Bai, D. Burek, and E. MacDonald. An integrated memory self test and EDA solution. In *IEEE International Workshop on Memory Technology, Design and Testing (MTDT'04)*, pages 92–95, August 2004.
72. F. Arnaud, F. Boeuf, F. Salvetti, D. Lenoble, F. Wacquant, et al. A functional $0.69\mu m^2$ embedded 6T-SRAM bit cell for 65nm CMOS platform. In *Symposium on VLSI Technology*, pages 55–56, 2003.
73. ITRS. International Technology Roadmap for Semiconductors (ITRS-2005); $http://www.itrs.net/common/2005itrs/home2005.htm$.
74. Ad J. van de Goor and I. Schanstra. Address and data scrambling: Causes and impact on memory tests. In *IEEE International Workshop on Electronic Design, Test and Applications*, pages 128–136, January 2002.
75. R. Dekker, F. Beenker, and L. Thijssen. Fault modeling and test algorithm development for static random access memories. In *IEEE International Test Conference*, pages 343–352, September 1988.
76. P. Shepard. Programmable built-in self-test method and controller for arrays. *US Patent No. 5,633,877*, May 1997.
77. J. Soden and C. Hawkins. I_{DDQ} testing: issues present and future. *IEEE Design and Test of Computers*, pages 61–65, October 1996.
78. S. Davidson. Is I_{DDQ} yield loss inevitable? In *IEEE International Test Conference*, pages 572–579, October 1994.
79. O. Semenov, A. Vassighi, M. Sachdev, A. Keshavarzi, and C. Hawkins. Effect of CMOS technology scaling on thermal management during burn-in. *IEEE Transactions on Semiconductor Manufacturing*, 16:686–695, November 2003.
80. T. M. Mak. Is CMOS more reliable with scaling? In *IEEE International On-Line Testing Workshop*, July 2002.
81. T. Kim, W. Kuo, and W. Chien. Burn-in effect on yield. *IEEE Transactions on Elecronic Package Manufacturing*, 23:293–299, October 2000.
82. A. Vassighi and M. Sachdev. *Thermal and Power Management of Integrated Circuits*. Springer, 2006.
83. A. Vassighi, O. Semenov, and M. Sachdev. Impact of power dissipation on burn-in test environment for sub-micron technologies. In *IEEE International Workshop Yield Optimization and Test*, July 2001.
84. J. Soden. I_{DDQ} testing for submicron CMOS IC technology qualification. In *IEEE Intenational Workshop on I_{DDQ} testing*, pages 52–56, November 1997.
85. C. Hawkins, J. Soden, A. Righter, and J. Ferguson. Defect classes - an overdue paradigm for CMOS IC testing. In *IEEE International Test Conference*, pages 413–425, October 1994.

86. T. Willams, R. Dennard, R. Kapur, M. Mercer, and W. Maly. I_{DDQ} test: Sensitivity analysis of scaling. In *IEEE International Test Conference*, pages 786–792, October 1996.
87. A. Keshavarzi, K. Roy, and C.F. Hawkins. Intrinsic leakage in deep submicron CMOS ICs measurement-based test solution. In *IEEE Transactions on VLSI Systems*, page 717–23, December 2000.
88. R. Dennard, F. Gaensslen, H. Yu, V. Rideout, E. Bassous, and A. LeBlanc. Design of ion-implanted MOSFET's with small physical dimensions. *IEEE Journal of Solid State Circuits (JSSC)*, SC-9:256–268, May 1974.
89. M. Sachdev. Deep submicron I_{DDQ} testing: Issues and solutions. In *European Design and Test Conference (ED&TC)*, pages 271–278, March 1997.
90. A. Keshavarzi, K. Roy, and C.F. Hawkins. Intrinsic leakage in low-power deep submicron CMOS ICs. In *IEEE International Test Conference*, pages 146–155, November 1997.
91. A. Gattiker and W. Maly. Current signatures: Applications. In *IEEE International Test Conference*, pages 156–165, November 1997.
92. P. Maxwell et al. Current ratios: A self-scaling technique for production I_{DDQ} testing. In *IEEE International Test Conference*, pages 1148–1156, October 2000.
93. C. Thibeault. An histogram-based procedure for current testing of active defects. In *IEEE International Test Conference*, pages 714–723, September 1999.
94. A. Miller. I_{DDQ} testing in deep submicron integrated circuits. In *IEEE International Test Conference*, pages 724–729, September 1999.
95. B. Kruseman, P. Janssen, and V. Zieren. Transient current testing of $0.25 - \mu m$ CMOS devices. In *IEEE International Test Conference*, pages 47–56, September 1999.
96. A. Keshavarzi, J. Tschanz, S. Narendra, K. Roy, C. Hawkins, W. Daasch, and M. Sachdev. Leakage and process variation effects in current testing on future CMOS circuits. *IEEE Design and Test of Computers*, pages 36–43, September 2002.
97. H. Kato, M. Matsui, K. Sato, H. Shibata, K. Hashimoto, T. Ootani, and K. Ochii. SRAM cell stability under the influence of parasitic resistances and data holding voltage as a stability prober. *IEEE Journal of Solid State Circuits (JSSC)*, 32:232–237, February 1997.
98. R. Rodriguez, J. Stathis, B. Linder, S. Kowalczyk, C. Chuang, R. Joshi, G. Northrop, K. Bernstein, A. Bhavnagarwala, and S. Lombardo. The impact of gate-oxide breakdown on SRAM stability. *IEEE Electron Device Letters*, 23:559–561, 2002.
99. A. Bhavnagarwala, S. Kosonocky, C. Radens, K. Stawiasz, R. Mann, Q. Ye, and K. Chin. Fluctuation limits and scaling opportunities for CMOS SRAM cells. In *IEEE International Electron Devices Meeting IEDM Technical Digest*, pages 659–662, December 2005.
100. A. Wang, C. Wu, R. Shiue, H. Huang, and K. Wu. New screen methodology for ultra thin gate oxide technology. In *IEEE 42-nd Annual Reliability Physics Symposium*, pages 659–660, April 2006.
101. F. Monsieur, E. Vincent, and D. Roy. A thorough investigation of progressive breakdown in ultra-thin oxides. Physical understanding and application for industrial reliabililty assessment. In *IEEE Proc. International Reliability Physics Symposium*, pages 45–54, April 2002.
102. S. Thompson, M. Alavi, M. Hussein, P. Jacob, C. Kenyon, P. Moon, M. Prince, S. Sivakumar, S. Tyagi, and M. Bohr. 130nm logic technology featuring 60nm transistors, low-k dielectrics and Cu interconnects. *Intel Technology Journal*, 6:5–13, June 2002.
103. S. Mukhopadhayay, H. Mahmoodi-Meimand, and K. Roy. Modeling and estimation of failure probability due to parameter variations in nano-scaled SRAMs for yield enhancements. In *Symposium On VLSI Circuits*, pages 64–67, November 2004.
104. M. Agostinelli, J. Hicks, J. Xu, B. Woolery, K. Mistry, K. Zhang, S. Jacobs, J. Jopling, W. Yang, B. Lee, T. Raz, M. Mehalel, P. Kolar, Y. Wang, J. Sandford, D. Pivin, C. Peterson, M. DiBattista, S. Pae, M. Jones, S. Johnson, and G. Subramanian. Erratic fluctuations of SRAM cache Vmin at the 90nm process technology node. In *IEEE International Electron Devices Meeting*, pages 655–658, December 2005.
105. J. Banik, A. Meixner, G. King, and D. Guddat. Static random access memory SRAM having weak write test circuit. *US Patent No. 5,559,745*, September 1996.
106. D. Weiss, J. Wuu, and R. Reidlinger. Integrated weak write test mode WWTM. *US Patent No. 6,192,001 B2*, February 2001.

107. William Schwarz. Data retention weak write circuit and method of using same. *US Patent No. 5,835,429*, November 1998.
108. Thomas Liston and Lawrence Herr. Method and apparatus for soft defect detection in a memory. *US Patent No. 6,590,818*, July 2003.
109. J. Yang, B. Wang, and A. Ivanov. Open defects detection within 6T SRAM cells using a no write recovery test mode. In *Proc. of IEEE 17th International Conference on VLSI Design*, pages 493–498, October 2004.
110. Roelof Herman Willem Salters. Device with integrated SRAM memory and method of testing such a device. *US Patent No. 6,757,205*, June 2004.
111. Moty Mehalel. Short write test mode for testing static memory cells. *US Patent No. 6,256,241 B1*, July 2001.
112. E. Selvin, A. Farhang, and D. Guddat. Programmable weak write test mode. *US Patent No. 6,778,450 B2*, August 2004.
113. K. Zhang, U. Bhattacharya, Z. Chen, F. Hamzaoglu, D. Murray, N. Vallepali, Y. Wang, B. Zheng, and M. Bohr. SRAM design on 65-nm CMOS technology with dynamic sleep transistor for leakage reduction. *IEEE Journal of Solid State Circuits (JSSC)*, 40:895–901, April 2005.
114. J. Wuu, B. Stackhouse, and D. Weiss. Programmable weak write test mode (pwwtm) bias generation having logic high output default mode. *US Patent No. 7,133,319 B2*, November 2006.
115. José Pineda de Gyvez, Manoj Sachdev, and Andrei Pavlov. Test for weak SRAM cells. *US Patent No. 7,200,057 B2*, April 2007.
116. A. Pavlov, M. Sachdev, and J. Pineda de Gyvez. Weak cell detection in deep-submicron SRAMs: A programmable detection technique. *IEEE Journal of Solid State Circuits (JSSC)*, 41:2334–2343, October 2006.
117. A. Pavlov, M. Azimane, J. Pineda de Gyvez, and M. Sachdev. Word line pulsing technique for stabilty fault detection in SRAM cells. In *IEEE International Test Conference (ITC)*, November 2005.
118. R. Baumann. Soft errors in advanced semiconductor devices – part i: The three radiation sources. *IEEE Transactions on Device and Materials Reliability*, 1:17–22, March 2001.
119. T. May and M. Woods. Alpha-particle-induced soft errors in dynamic memories. *IEEE Transactions on Electron Devices*, 26:2–9, January 1979.
120. L. Massengill. Cosmic and terrestrial single-event radiation effects in dynamic random access memories. *IEEE Transactions on Nuclear Science*, 43:576–593, April 1996.
121. R. Baumann. Radiation-induced soft errors in advanced semiconductor technologies. *IEEE Transactions on Device and Materials Reliability*, 5:305–316, September 2005.
122. R. Baumann. Technology scaling trends and accelerated testing for soft errors in commercial silicon devices. In *IEEE International On-line Testing Symposium IOLTS*, page 4, 2003.
123. R. Baumann. Soft errors in advanced computer systems. *IEEE Design and Test of Computers*, 22:258–266, May 2005.
124. P. Roche et al. Determination of key parameters for SEU occurrence using 3-D full cell SRAM simulations. *IEEE Transactions on Nuclear Science*, 46:1354–1362, December 1999.
125. R. Baumann. Ghost in the machine: A tutorial on single-event upsets in advanced commercial silicon technology. In *a tutorial at IEEE International Test Conference (ITC)*, November 2004.
126. V. Degalahal, N. Vijaykrishnan, and M. Irwin. Analyzing soft errors in leakage optimized sram design. In *IEEE International Conference on VSLI Degisn*, pages 227–233, January 2003.
127. P. Dodd and F. Sexton. Critical charge concepts for CMOS SRAMs. *IEEE Transactions on Nuclear Science*, 42:1764–1771, December 1995.
128. R. Baumann and E. Smith. Neutron-induced boron fission as a major source of soft errors in deep submicron SRAM devices. In *Proceedings of IEEE International Reliability Physics Symposium*, pages 152–157, April 2000.

129. Y. TOsaka, H. Kanata, T. Itakura, and S. Satoh. Simulation technologies for cosmic ray neutron-induced soft errors: Models and simulation systems. *IEEE Transactions on Nuclear Science*, 46:774–779, June 1999.
130. T. Karnik, P. Hazucha, and J. Patel. Characterization of soft errors caused by single event upsets in CMOS processes. *IEEE Transactions on Dependable and Secure Computing*, 1:128–143, April 2004.
131. Hamming Code. Wikipedia, *http://en.wikipedia.org/wiki/hamming_code.*
132. D. Bhattacharryya and S. Nandi. An efficient class of SEC-DED-AUED codes. In *Third International Symposium on Parallel Architectures, Algorithms and Networks*, pages 410–416, October 1997.
133. D. MacKay. *Information Theory, Inference and Learning Algorithms*. Cambridge University Press, ISBN 0-521-64298-1, 2003.
134. R. Baumann. The impact of technology scaling on soft error rate performance and limits to the efficacy of error correction. In *IEEE Electron Devices Meeting*, pages 329–332, October 2002.
135. J. Maiz, S. Hareland, K. Zhang, and P. Armstrong. Characterization of multi-bit soft error events in advanced SRAMs. In *IEEE International Electron Devices Meeting*, pages 519–522, December 2003.
136. V. Degalahal, L. Li, V. Narayanan, M. Kandemir, and M. Irwin. Soft error issues in low-power caches. *IEEE Transactions on Very Large Scale Integration (VLSI) Systems*, 13:1157–1166, October 2005.
137. N. Seifert, D. Moyer, N. Leland, and R. Hokinson. Historical trends in alpha-particle induced soft error rates of the $Alpha^{TM}$ microprocessor. In *IEEE International Reliabilty Physics Symposium*, pages 259–265, April 2000.
138. H. Nguyen and Y. Yagil. A systematic approach to SER estimation and solutions. In *IEEE International Reliabilty Physics Symposium*, pages 60–70, March 2003.
139. X. Zhu, R. Baumann, C. Pilch, J. Zhou, J. Jones, and C. Cirba. Comparison of product failure rate to the component soft error rates in a multi-core digital signal processor. In *IEEE Annual International Reliability Physics Symposium*, pages 209–214, April 2005.
140. P. Shivakumar, M. Kistler, S. Keckler, D. Burger, and L. Alvisi. Modeling the effect of technology trends on the soft error rate of combinational logic. In *IEEE International Conference on Dependable Systems and Networks*, pages 389–398, June 2002.
141. Moty Mehalel. Lowering voltage for cache memory operation. *US Patent Appl. No. US 2006/0161831 A1*, July 2006.
142. C. Chen and A. Somani. Fault-containment in cache memories for TMR redundant processor systems. *IEEE Transactions on Computers*, 48:386–397, April 1999.
143. N. Calin, M. Nicolaidis, and R. Velazco. Upset hardened memory design for submicron CMOS technology. *IEEE Transactions on Nuclear Science*, 43:2874–2878, December 1996.
144. J. Liaw and Hsin-Chu. SRAM cell for sort-error rate reduction and cell stability improvement. *Patent Application No. US 2005/0265070 A1*, December 2005.
145. R. Wilson. ST tames soft errors in SRAM by adding capacitors. EE times Jan. 2004l, *http://www.eetimes.com/showarticle.jhtml?articleid* = 18310682.
146. Actel. IGLOO low-power flash FPGAs with flash freeze technology; *http://www.actel.com/documents/IGLOO_DS.pdf.*
147. J. McCollum. Radiation tolerant sram bit. *Patent Application No. US2005/0193255 A1*, September 2005.
148. P. Hazucha and C. Svensson. Impact of CMOS technology scaling on the atmospheric neutron soft error rate. *IEEE Transactions on Nuclear Science*, 47:2586–2594, December 2000.
149. A. Agarwal, H. Li, and K. Roy. A single-v_{TH} low-leakage gated-ground cache for deep submicron. *IEEE Journal of Solid State Circuits (JSSC)*, 38:319–328, February 2003.
150. A. Nourivand, C. Wang, and M. Ahmad. An adaptive sleep transistor biasing scheme for low leakage SRAM. In *IEEE International Symposium on Circuits and Systems ISCAS*, pages 2790–2793, May 2007.

Index